RANGER'22

Ray Goggins served in the Irish Army for 26 years, including 17 years in the Army Ranger Wing as an operator and leader in a Tier 1 Special Operations Unit. With a lifetime of experience in war zones, Ray has also operated as a bodyguard in Europe, Eurasia, the Middle East and Afghanistan. Born in Cork, Ray is now chief instructor on RTÉ's *Ultimate Hell Week* and director of a training and leadership company, Coreskill, working with various corporate organisations and teams.

RANGER22

LESSONS FROM THE FRONT

RAY GOGGINS

Gill Books

Gill Books
Hume Avenue
Park West
Dublin 12
www.gillbooks.ie

Gill Books is an imprint of M.H. Gill and Co.

© Ray Goggins 2021
9780717192496

Design and print origination by O'K Graphic Design, Dublin
Edited by Djinn von Noorden
Proofread by Susan McKeever
Printed by CPI Group (UK) Ltd, Croydon, CR0 4YY

This book is typeset in 12/18 pt Minion.
The paper used in this book comes from the wood pulp of managed forests. For every tree felled, at least one tree is planted, thereby renewing natural resources.

5

This book is dedicated to those who have stood on the front line to protect others, mostly without anybody knowing about it.

And

Especially to those who didn't make it home, particularly John and Derec, who would have written much better books.

———◆———

Glaine ár gcroí, neart ár ngéag, agus beart de réir ár mbriathar

Cleanliness of our hearts, strength of our limbs and commitment to our promise

CONTENTS

INTRODUCTION

My name is Ray Goggins. I have spent all of my adult life in the military or in a parallel line of work. Most people know me as the chief instructor on RTÉ's *Special Forces: Ultimate Hell Week*. It has taken me 35 years to become involved in that *Special Forces* programme – being on TV was never on the radar, or something I even thought about. I had previously served in the infantry, but mainly in Special Forces as an operator and leader in hostile environments, conflict areas and war zones all over the world. A lot of things happened along the way, which taught me the skills and attitude to become that person on TV.

Most people think that Special Forces units are solely in the business of taking people out as a last resort. Sometimes this is the case – these units operate between the spaces that conventional units can't fill. Sometimes the mission is too complex for conventional forces or may just require a small team of highly skilled individuals.

In fact Special Forces (also known as Special Operations) units are there to protect people from terrorist and criminal organisations. I have spent most of my career in this capacity, from protection operations in Afghanistan to conducting direct-action (DA) missions in Africa, Asia and the Middle East. The greatest reward is making that difference for people, physically saving them from injury and death. By carrying out operations we can ensure that the village they live in is not attacked by insurgents, that they can harvest their crops safely, and that their children are not taken from them to become slaves or child soldiers.

Let me state for the record that I'm no psychologist, therapist, or life coach, yet at some stage or other I've been all of the above, both for myself and others. I don't have all the answers, but what I do have is a tried-and-tested process, which I have developed in the military and beyond over many years of being caught up in hostile and dangerous situations. The process is based on everything from doctrine and drills to practical crisis-management experiences, which I have used in my military and civilian life.

The military has the best guidelines in that a manual, pamphlet, training circular, routine order, screed or a list of instructions exists for everything – from how to tie your shoe laces to how to drive a diver propulsion device (DPD). These guidelines are called standard operating procedures (SOPs) – and if there isn't an SOP for it, then it doesn't exist!

So much of what I have learnt in the military can be transferred to everyday life and its many challenges, until it becomes second nature. And because life doesn't have an SOP, I have developed my own.

———◆———

Excluding my Special Operations career, I have worked in private security as a manager and director, at corporate level, and as a consultant private contractor in the Middle East and Central Asia. I've been a bodyguard and security protection officer to the rich and famous, a physical trainer, instructor, teacher and mentor and

I even worked in the fashion tech industry (still not sure how that happened!).

Special Operations is not just about the hard skills and sharp-end abilities. The by-product of all that high-octane action and training is that you also learn other, equally important 'soft' skills, or an ability to problem-solve in a non-kinetic (in other words, non-physical), or non-violent manner. Intelligence operations and protection-related missions, for example, require a much higher standard of operator – not just an assault moron who presses the trigger to score a body count.

In this book I want to show you how a particular mind-set can help you manage life's ups and downs. I'll also share some insight into my own experiences, which show how they worked for me – or didn't, in some cases. I'll demonstrate some simple techniques, drills and lessons to help you navigate life and maybe make better decisions: and if it does nothing else, this book might make you feel better about your life choices when you read about some of the stupid ones I've made in mine!

This isn't about me telling you how fantastic I am and all the great stuff I have achieved and done – well maybe a little (I'm joking). The path I have travelled, the things I have seen and done and the people I have met in crisis situations have given me a window into those qualities that make us tick effectively. My military training has defined me over many years and created a mind-set, outlook and skills that can be trained and channelled into various situations.

―◆―

I will always think as a soldier. It's my factory setting ('factory setting' is something I picked up in the military: when you get a new piece of kit or equipment it comes 'factory-set' to operate a particular way within certain parameters). Our factory setting is our core values; how we think, behave and react. It colours how we perceive and live our life, how we process information, and how we react and relate to others in situations we and they create.

In the main we just get along with our day-to-day lives, but our factory setting begins to affect us for better or worse when a benign situation deteriorates and develops into a crisis. When the situation goes south is when we get to the real person. It's when the bullshit disappears. Under pressure, we revert to type – our factory setting – and follow that behavioural process ingrained in us, be it good or bad. The stress is not the event itself but your perception and reaction to it, and this can be trained and managed. You have a far more positive effect on a situation when you are focused and remain calm, composed and in control of your own actions first and foremost.

I wrote this book to explain how I started on this road, and how I collected the qualities I believe are critical for me and how I function. Dealing with a truckload of armed militia in Africa and trying to get a group of school children to follow directions, although very different, are both situations that carry the same basic principles.

I have learnt how to deal with my own fears of life, the ones we all have, and how to react and behave as a result. Our brains are not actually wired for success or risk – in fact they are completely focused on keeping us safe and making sure we survive. Any thoughts or actions we may have outside that zone of survival, the brain will quickly want to stop. It considers these a threat, whether they really are or not. Sometimes you have to bypass the brain asking you what you're doing in order to step outside the safe zone.

I have been blessed with great people along the way who have shown me the ropes, picked me up, covered my ass and kicked it when I deserved it. This extends to a great supportive family, good friends and some absolutely amazing workmates. That bond is critical with those around you in the military – and in Special Forces in particular, where your team is your family. I am incredibly lucky to have great friendships with guys who are more like brothers – we have been through hell together, both in training and on operations, and this has fused us together.

To protect my former unit and team mates I must follow protocols that I am bound to observe by a code of honour and by military law. This information is restricted in order to protect materials, names, locations and operations in which I may have been involved.

———◆———

So how did a book come about if I spent a lifetime under the radar, maintaining operational security?

It all started when I decided I wanted to serve in Special Forces and applied to undergo selection for the Irish Army Ranger Wing (ARW) – or to give it its proper title, *Sciathán Fianóglach an Airm*. During selection candidates are tested in every physical, mental, emotional and psychological way in order to identify individuals who are deemed potential recruits with the core values required for the ARW. The course is incredibly tough: the failure rate has been 100 per cent some years, but in general it is around 90 per cent.

On passing this course the candidate will progress onto a skills course, run over several months, and those successful in passing will be taken into the unit as operators, serving a probationary period for 18 months. This is the first rung on the ladder and the beginning of an education and immersion into the Special Forces world.

Being a Ranger is not a job or profession – it is a transition into a different way of life and state of mind, which is honed over years and requires you to commit to it, body, mind and soul. At times you will forsake all others, not least your own family and friends, to serve the idea of a belonging to something that is greater than you are. You forego promotion and a regular army career path to serve in a unit where you give everything for a reward that can't really be seen in your bank account or on your record. You don't get a pat on the back, but the reward is immeasurable. All your effort, ability and will to succeed in all situations to achieve the mission is taken as the only norm. The people around you are why you do it.

◆

I was lucky to serve in the ARW for 17 years, where I was deployed to various hot spots around the globe and carried out operations and missions, both at home and abroad. When not deployed or on operation, the unit was on a punishing training regime to keep standards as high as possible and technical skills up to date. Physical fitness, mental agility and technical ability were the main categories: all other skills fell into those categories.

After a number of years of this I moved on and retired from the ARW to take up a position as a consultant protection officer for a global company. I was happy to leave the army on my own terms and in a good place in my relationship with it, but I just wanted to do something a little different before I got too old, I guess. The army and ARW made me who I am today and I am always thankful and proud to be associated with it. It's a bit ironic, because when I served in the ARW I couldn't tell anyone where I worked – and now I'm on TV and have written a book!

I loved my time spent serving in the army and in Special Forces in particular and I will always be thankful for the life it showed me. It allowed me to learn and see things that I otherwise would never have even imagined, never mind experienced. Yet in the end, I left the army when I did because I wanted change, and I was young enough to start a new chapter in my life. The pay also was a problem – allowance cuts and pay freezes had made things barely manageable and not enough for the effort put in so that pushed my decision to jump ship to the private sector.

Leaving the army was hard but it gave me the freedom to try new things, such as eventually running *Special Forces: Ultimate Hell Week*, the TV show in which we put civilian recruits through Special Forces training. It turned out to be hugely popular, and after a flurry of media write-ups and interviews we were very much out of the closet, so to speak, so much so that after Series 2 I was contacted by Gill Books. Would I be interested in writing down some of my experiences and sharing my life story? This had never been done before because of course the ARW guys operate anonymously.

So that's the short version of how I got to this stage. I've made mistakes and learnt some hard lessons. And despite all those years of breathing pressurised oxygen underwater, taking Lariam anti-malarial tablets in Africa and too many concussions, I still remember the stuff that counts. I hope you can take something from it as this is how it happened.

PROLOGUE

A biting cold wind came off the Hindu Kush that January evening. It was shortly before 7 p.m. on 14 January 2019. With the temperature well below zero, a light blanket of snow shrouded the compound. I stood just outside the door of my room in the long reinforced corridor of our accommodation block, situated in a multinational secure compound just off the Jalalabad road in Kabul, Afghanistan.

I was waiting for my bosses and good friends Davy and Vincent to emerge from their rooms so we could go and get dinner in the DFAC (dining facility). I've never been good at waiting, and was even starting to get a little hangry. Just as I was about to knock for the guys to get a move on, a sudden huge explosion lifted me off my feet and threw me through the door back into my room, where I bounced off the wall and landed on the floor.

The sound was instant. It was on me a millisecond after I felt the first inclination of something, like a sixth sense of impending doom, if you will. (Working in a combat zone quickly equips you with near-clairvoyant abilities at times.) The blast felt like someone had picked me up in a bear hug and pushed me through the door. It assaulted all my senses at once, sight immediately into darkness, sound of a dull but heavy blast and the feeling of being lifted up despite myself. Luckily it was a light door so I didn't feel the impact until I hit the concrete wall and then the floor. Then silence and a buzzing in my ears.

'Fuck!' I said out loud. I lay on my back for a few seconds before my brain began rebooting, kickstarting into action as the impact began to wear off. I made a quick check of my body for injury. I wasn't wearing any protective kit as we were dressed casually in the compound; I just wore a CZ automatic pistol in a covert holster. I felt like I had been kicked in the chest; my ears were ringing and I could feel a slight sting on my head as a trickle of blood ran down my forehead, but I thought I was good so I stood up.

My immediate action was to check my pistol and get my main weapon, an AK–47 assault rifle, into my hands, get my kit on and go deal with whatever was coming next. I was a private security contractor responsible for the safety and protection of people working in one of Afghanistan's largest telecom companies. I knew from experience that those who had blown the initial vehicle-borne improvised explosive device (VBIED) would be following up with a ground assault in suicide vests and coming in to kill as many people as they could. I quickly put on my body armour, did a press check on both weapons to be sure there was a round chambered, checked my spare magazines and stepped out the door into the darkness.

CHAPTER 1

BRED IN THE BONE

'With the honesty of purpose, balance, a respect for
tradition, courage, and, above all, a philosophy of life, any
young person who embraces the historical profession will
find it rich in rewards and durable in satisfaction.'

SAMUEL ELIOT MORISON

My family has a significant history of long service in the army, first the British Army and then the Irish Army once it was up and running, a lineage handed down from father to son over a number of generations. And yet much of the information of this historical military vocation only came to light later on in my life – it seems that it was very quietly handed down. Silent leadership was obviously my family's style. My own father never pushed me to join the army and rarely spoke of it, to be honest. It just seems that it was genetically implanted in our DNA.

It all began in 1854 when my great-grandfather, Martin Goggins, joined the Royal Artillery in Manchester. Born in 1833 in the parish of Moneygall in the King's County (now Offaly), he had survived the Famine. The British Army keep very factual and dull records, but I found out that he had been just 22 years old when he joined up. His occupation was listed as gardener. He was described as having a fair complexion, dark hair and green eyes (sounds like my enlistment description 136 years later!).

Martin was promoted to bombardier and corporal later on in his career, travelling as far afield as Gibraltar but mainly serving in the British Isles. He married in 1866 and his son James, my grandfather, was born in married quarters on Haulbowline Island in Cork Harbour in 1878. Martin retired to Cobh as a military pensioner and returned to his original profession as a gardener until his death in 1902.

My grandfather, James, joined the British Army in 1898 in Haulbowline at the age of 19. He joined the Royal Artillery like his father, going on to serve all over the world. At one stage he was posted to Mauritius in a garrison battery (a group of troops stationed in a fortress or town), but barrack life seems to have taken its toll on him and his disciplinary record wasn't the best. He was up on offences of gambling, drunken disorder and insubordination – the usual stuff for an Irishman in the British Army when there is no war on. He deserted the army sometime early in the new century and the trail goes cold for a while.

It doesn't end there, though, as James resurfaces as a serving British soldier on the Western Front in 1915 in a medium trench mortar battery attached to the 5th Division, in the second Battle of Ypres. In March 1916 his battery was in the line on the southern edge of Vimy Ridge in front of Arras and, following a period of rest, James and his division were put into the slaughter of the Somme in July, fighting in some of the major battles.

James and his comrades were again redirected in the winter of 1917 to fight in Italy, to support the Italians who needed help in the north of the country. The division ended up along the River Piave in January 1918 and saw battle once again, where he was individually decorated for courage in action. The Italy campaign finished up soon enough and James was back on the Western Front and was more or less in action all the way, fighting over the Somme once again and on to the war's end and the division demobilisation in December 1918.

My grandfather returned home sometime in 1919 to a different Ireland from the one he had left. He was obviously changed from years in the trenches and had to adapt quickly on his return, partly because Irish war veterans were not welcomed home at the time. The political situation had changed, as had sentiments towards the British Army now involved in a war against Irish nationalists fighting for independence. People now felt contempt for all things associated with the Crown, including World War I veterans.

James married in the 1920s and settled down in Ballyhooly in north Cork. He never told a soul about his experiences in the army and put his medals away, never to see the light of day. This was the norm at the time for a lot of returning Irish men, in view of the political and civil situation.

I have just one handed-down story, told to me by my uncle, of James as part of a work gang digging drainage ditches in the Fermoy area in the early 1920s. Himself and another war veteran were in a group of six men, when the other four put it up to the two veterans about their service with the British Army. It got violent and the other navvies attacked the ex-soldiers to teach them a lesson for taking the King's shilling. Fortunately for James and his descendants, namely me, the lesson went the other way as the two men had spent five years fighting in trenches and this was just another skirmish to them. Apparently, James left the county council the same day (or was sacked, more likely) and took up a new job as a caretaker in what was always known as 'the Mansion' in Ballyhooly, a place I knew well from spending summers there as a boy. It came with a caretaker's cottage on site so James moved in and started his family, and my father Daniel was born there.

You might know the big house by its now more popular name of Castlehyde.

My grandfather died in 1951 and was laid to rest in Rafeen, back in Cork Harbour, along with his stories and experiences of the war to end all wars. He died 20 years before my birth and my father never mentioned him. I'm not particularly religious and I don't believe we get to meet all those who went before us but if I could sit down for a quiet pint with someone from the past and hear their story, I would love it to be him.

———◆———

Next up for the family business was my father Daniel who, having been born in the surrounds of what would become Michael Flatley's mansion, was off to a good start. He finished school and worked as a farm labourer for some time before he went and joined the 4th Infantry Battalion in Collins Barracks in Cork City, shortly after his father's death in 1951. He was 17 years old at the time and I think he was influenced by his own father and what was now becoming a family tradition of enlisting in the army.

The Irish Army at the time was not as operational as it would become. There were no overseas deployments and training concentrated more on domestic soldierly duties. My dad spent some years at this, amidst constant inspections and regimental parades, 'bulling' boots and shining brass buttons. This he carried for life into retirement. I can remember as a child, on a Sunday morning, the smell of Kiwi Shoe Polish as my dad would shine all the shoes in the house prior to the Goggins family parading for mass. He could also put a crease in your trousers using brown paper that was sharp enough to shave with.

Daniel made corporal by 1959 and in 1960 commanded a medium machine gun (MMG) section with B Company 32nd Infantry Battalion, formed to serve in the Congo in Central Africa – the first ever deployment of Irish troops on a United Nations mission in the history of the state. This would set the conditions and tradition for Irish troops to serve on United Nations missions with distinction and valour for generations to come and bring peace to countries in every corner of the planet, at no small cost in lives and injuries.

The Congo had recently been given independence from Belgium, which was still intervening with Belgian troops and mercenaries in that vast country, causing widespread instability and violence, particularly in the mineral-rich Katanga province in the south of the country. Katanga was infamous for having supplied the uranium used in the atomic bombs dropped on Japan at the end of World War II.

My dad never told me any war stories about the Congo, or anywhere else for that matter, but preferred to talk about the day-to-day living and of course some of the craic he had in Africa, like keeping a pet monkey with his unit. He did, however, tell me a story about the three-day train journey from Kindu to Kamina, which was the nearest thing to a war story I was ever going to get.

The journey took some 76 hours on a train consisting of mainly flat cars loaded with provisions and equipment and a covered car at the end, which served as a strong point and command centre. As part of the MMG section, my dad had a Vickers machine gun set up in front of the engine that covered the axis of travel. Up ahead they noticed that approximately 300 yards of railway track had been removed; he gave the signal and with a screeching of brakes the train slowly ground to a halt in a deep cutting with the troops looking up at the jungle above on both sides and an ambush they couldn't have foreseen. They found themselves surrounded by several hundred Baluba. The Baluba was an indigenous tribe with a reputation of hostilities towards Europeans in the region. They were armed with everything from rifles and spears to bows and arrows and also had their women and children lined up in order of battle.

It's no coincidence that Irish troops (nowadays globally revered as UN troops), with no colonial hangovers, have earned a reputation for defusing situations without bloodshed, and this was one of the instances where this principle was borne out. My dad said they could see that a lot of these people were starving and sick, and so a dialogue of sorts was organised, using an interpreter, to conclude the ambush by simply giving the Baluba food, as several sacks of grain were stored on the flat cars. The soldiers even managed to get the Baluba to repair the track before they set off to complete the final leg of the journey.

Daniel – or 'Danny' as my dad was known in the army – went on to serve again in the Congo on a second deployment as sergeant and was then posted to Cyprus on four occasions during the United Nations operation there at a time of insurgency and conflict in the sixties. He would also be part of a famous guard of honour for President John F. Kennedy during the latter's visit to Ireland in 1963, something he was hugely proud of all his life. More difficult for him was his experience dealing with the displaced during internment and the difficult times the country had in a turbulent 1969, when he was part of a unit that assisted interned people and their families who were housed in an army camp in Kilworth.

I don't remember my dad in the army. I was too young, just two, when he retired from service in 1973 after an exemplary career. He was always in good form and full of humour at home. I'm sure he had horror stories as well, but even when I experienced my own, we never shared them. This trait I still have now with my own family as, like my dad and Irish lads in the main, we just talk about the good stuff and leave the bad stuff out. Without him ever knowing it, I think he was the greatest influence on me joining up, not because he ever particularly encouraged me to do so, but because he was so calm and easy-going about it all.

◆

As they say in the army, 'Next man!' which happened to be James (Jim), my older brother. Jim joined up in 1979 and served in the 4th Infantry Battalion from the outset of his long career. He served overseas in Lebanon on United Nations peacekeeping missions on three separate occasions in different battalions in the 1980s and 90s.

Jim was also a massive influence on my choice of occupation as I could see some of his life in the army at second hand. I can vividly remember going to his passing-out (graduation) parade in 1979, which left a huge impression on me. I must have been just eight or nine at the time but I can remember in detail the parade and the arms drill of his platoon carrying FN rifles – my first view of a weapon, which has always stayed with me. I couldn't pick out the individual soldiers in the platoon: the coordination was perfect and every single man was a part of this one movement. It was also my first experience of an army stand-up buffet lunch. I wondered why we had to stand at a perfectly dressed table with individual place settings but no chairs!

I didn't know it at the time, but this was the start of my military education. I was assimilating my first and most important factory setting. It is probably the most basic and important core value that I have had all my life of what the army should be: a group of people working together as one, with intent and purpose, to achieve the immediate aim.

When Jim was overseas he would send me letters, photos and tapes from Lebanon. I would stick the pictures on my wall and inhale every word of every letter. He sent photos of himself on patrol or manning a machine-gun position surrounded in sandbags, which made a huge impression on 14-year-old me. Later, after I joined up, I was lucky to serve in the same company with Jim for close to nine years.

We served together on several regimental duties and operations in Ireland, but were never deployed overseas together. The closest we got to that was when we met on the tarmac in Ben Gurion airport in Tel Aviv in 1993. I was deploying to Lebanon that summer and was disembarking from the aircraft Jim was about to board to go home to Ireland. The Military Police (MP) normally separate the two different units and as we filed past one another, the odd charming comment or wish was verbalised: 'Up your Swiss!' or 'Only six months to go, lads!' The troops also look different:

the guys coming home are in UN 'whites', a smart uniform of sand-coloured trousers and shirt, with blue UN berets, cravats and suntans. The new guys are in army green and are pasty pale white and on a downer as they are probably still hungover from the going-away party the night before.

Luckily there was a bit of a delay and during a pause in the movement I saw a figure in whites, stalking around the wing of the aircraft and heading in my direction. I immediately copped who it was and as some MP corporal bawled at me to get back into formation, I bolted. We ran towards each other and embraced on the scorching, melting tar of the runway in the open space between each column as a cheer went up from both battalions. It was a magic moment and Jim – being my big brother – handed me a wad of US dollars. I, on the other hand, gave him about three pounds and twenty pence. We laughed, said our goodbyes and went our separate ways, this fleeting view being the last time we would see one other for a year.

Jim served a total of 39 years of duty and service to an army he loved. He had followed a legacy begun by our forebears. Like my father, he always had a good word for all and was respected and honoured by generals and privates alike throughout the forces. I went to his retirement celebration in Collins Barracks and it was a testament to my big brother's good nature that the turnout of serving and ex-soldiers has rarely been matched since.

◆

Let me balance the story a little and talk about my mother's side of the family, the Dooleys. My great-grandfather Tom, though not a military man, was a devout GAA man. He was involved in the setting up of the organisation, particularly in Cork in the 1800s. He was part of the founding group of St Finbarr's GAA club in Cork and chairman of the county board in the early part of the twentieth century.

His son Patrick, my grandfather, was born in 1900 and was a fervent nationalist, taking part in the limited 1916 Rising in the Cork area at the age of 16. I am not really sure what he did during the War of Independence, but I do know that he was involved in some regard as an active member of the independence movement. As an older man he joined the Free State Irish Army in 1940 and served during the Emergency up until 1945, probably more as a reservist. He, like his father, was a passionate hurling man, which made its way down the line to my mam and to my siblings also.

So my military inheritance and influence came mainly from my paternal family – the British Army and what that brings – with the nationalist and old IRA side coming from my maternal ancestors. A good mix, I think: I'm proud of both histories and glad to be a descendant from both lineages, despite their political differences.

The person who has had a considerable influence on me is my oldest brother, Ger Dooley, with his sensible yet critical, no-nonsense approach to everything. Ger was ever straight and true and a great man for advice, along with his wife Mary, who was more like a sister to me. Ger was born before my mam met my dad and lived with my maternal grandparents in Bishopstown, to his betterment, he would say. Though he never served in the military, he had an interesting life and would have been a brilliant officer: he was so posh, he even had the trench coat! Ger lost his life to cancer in October 2019, having tragically lost Mary in 2016. We held his funeral the day before I started on Series 2 of *Hell Week*.

Ger firmly believed that what is bred into you comes out, regardless of what happens, and that you are a product of your genetics and what came before you. He always said, 'What's bred in the bone comes out in the marrow.'

———◆———

I was born in 1971, the youngest in a family of seven children. It's a large family by today's standard but on the north side of Cork City back then it was about average. We lived in a corporation house in Fair Hill. Our house was always busy with people coming and going. It was a wonderful childhood, happy, close and full of love.

Life for us was usually a laugh, although there was murder at times also, of course, in a small three-bed house with four girls, Ber, Anne, Catherine and Paula in one room and three boys, Jim, Dan and I in the other, a tiny box room. My family was fixated on sports and sporting activity, which I really didn't pick up until later, once I'd left home. In the house it was mainly hurling and soccer with a bit of track and field and whatever sport was on TV was re-enacted by us at home. My sisters excelled at sport, soccer and camogie, going on to win senior county all-Irelands and even representing Ireland in soccer at senior international level.

I was more of a movie fan, however, and while my family watched sports on a Saturday afternoon, I was often in the bedroom watching old matinees on a black-and-white portable TV instead. This is where I developed a love for old movies, which I have to this day, and where my view of life was shaped by John Wayne, Robert Mitchum and others.

My mam, Joan, as the family matriarch, carried out the disciplinary actions. My dad was a lot calmer and, having been a UN soldier, didn't use force on us. My mam, on the other hand, was fast and accurate with a slipper or the wooden baking spoon and would use it as required. She would also deny herself everything to get us what we needed and had an amazing work ethic. She would protect us like a lioness, something she did till her last breath.

I remember thinking I was great at hiding wooden spoons under the fridge over a period of probably years. That was all well and good until we got a new fridge and she wondered why I was hovering like a hen before my dad lifted up the old one. Of course there were four wooden spoons there for her to see and she laughed heartily, but I still got a slap of each one of them before they were returned to the cutlery drawer (or weapons locker).

My mam had a hard start in life. In 1951, after a previous relationship, she gave birth to my brother Ger when she was only 20 years old. Her parents sent her to Bessborough House, the infamous mother and baby home in Cork, which was more like a prison run by heartless nuns and where she was subject to appalling degradations and hardships. She never spoke of her time in the home, though her parents took Ger out of there soon after his birth to raise him themselves. They left my mam in there for some time before she was allowed home to a life of harsh treatment meted out by her own family. Despite the way she was treated by the Church she remained, throughout her life, a devout Catholic, and worked constantly to help others and the Church.

◆

As I said before, my upbringing was very stable. Humour was a constant in the home and it's a quality that I hold crucial and on which I have come to rely in life. Family support and team effort were also characteristics I learnt growing up: we all were shown how you unite and back each other up, something we still do as a family. Both my parents instilled respect and honesty in us and taught us that no matter what you did, if you stood up and told the truth you would only get half the 'bate-in' (beating), as we say in Cork. I have to admit I was on the receiving end a few times, which I fully deserved, and it did me no harm.

I had my mam's heart broken from an early age. I was a bit of a daredevil and constantly getting busted up in some way. I remember getting clipped by a car at four; at six I was hanging off the back of a neighbour's bread van when the door opened and I fell and was dragged behind the van for a bit. Mam picked me up and basically ran, carrying me in her arms, to the nearby North Infirmary. I broke a couple of bones from time to time and was a terror for creating realistic battles by burning plastic soldiers in the house, which she used to go mad over.

My dad had to dive in and save me at the age of three in Butlin's where we holidayed every summer after I jumped in at the deep end of the swimming pool and of course sank straight to the bottom. There were glass windows under the water looking out onto the street below so people could look in and see the rescue as it happened. It didn't have any adverse effect on me and I'm told I was straight back in the water as soon as I could. When I did learn to swim, it became a huge part of my life as we lived literally five minutes' walk from the local corporation pool in Churchfield. The staff knew me so well they just let me in without the ten-pence fee and I could stay as long as they were open. I would spend hours there, and always felt comfortable and at home in the water. I remember being there once when my shoes were stolen while I was swimming – my mam had just bought them for school, so my brother Dan had to give me a piggyback all the way home.

The street we lived on had a full-sized soccer pitch in the grass square so the place was packed with children playing and getting up to mischief. With such a varied group of ages and sizes, fighting and bullying were all part of the norm, as it was in school. I don't remember any stand-out example of being bullied but it was something I learnt to accept. Dan covered my ass a few times. He also occasionally gave me the digs himself, although nobody else outside the family was allowed to. As the next one up from me in the pecking order it was his birthright to be able to punch me. It was never heavy, just the odd slap, and as brothers just a few years apart we were always at it. It was a more innocent time then, and we would be out of the house from morning until night getting up to all sorts.

At primary school in Fair Hill, the principal would make his rounds of the classes every morning and carry out the main punishments of the day. It was rare that I was caught up in this but on the odd occasion he called out my name it left its mark in more ways than one. You would be brought to the top of the class with the other condemned and line up at the blackboard. The severity of your crime would decide how many slaps you got from the drumstick he carried in the inside pocket of his sports coat.

We were told to put out our hand, palm up, and he would adjust the height of the drop depending on the crime. When the blow was struck the pain was intense and the stinging on my hand lasted for several minutes as I choked back tears. I refused to give him, or the class, the satisfaction of seeing me cry. I got two slaps and returned to my seat where I put my hands straight on the underside of the desk, which was metal and would cool down my stinging flesh. And so I developed an understanding of violence from an early age. Although I tried to avoid it and never forced it onto others, when it came I accepted it, giving it back if I could. This experience taught me discipline – not in the traditional sense of being disciplined or getting a slap for mistakes, but to have the discipline to do something properly when I have the chance to.

One of my neighbourhood friends had a father who did some hunting and kept a .22 calibre rifle in his shed. This led to myself and a couple of friends stealing ammunition, and although we tried to get the rifle as well it didn't work out – probably for the best, since the three of us were only about nine or ten years old.

The plan was to light a small fire on the far side of the football pitch, out of sight from our houses. We had seen a Western where a cowboy had dropped a handful of bullets into a fire and this was something we were all dying to try out for ourselves. So we threw one or two rounds of the liberated .22 calibre ammunition into the fire and decided that we would take turns jumping over it as a test of manhood, if you like, not realising that we were exposing our manhood to the perils of gunfire.

We decided we would up the ante a little and drop in another handful of rounds and see what happened. As we did it another boy, a neighbour who was an altar boy of a similar age, passed by. He stopped to talk to us, shopping bag in his hand, and my two amigos started talking to him as I continued to jump over the fire. Suddenly there was a series of bangs and we all dived for cover in the crossfire as bullets flew everywhere.

We looked up just as the altar boy fell backwards and lay motionless on the ground, still holding the shopping bag. We immediately thought we'd shot him dead. Panic ensued, the only answer for which was to run away, which we did at once (it was now becoming a proper Western with three fugitives on the run). But the gravity of the situation had sunk in by the time we got to the next terrace, and we decided we should turn back.

We returned to the scene to find the body had disappeared, but as we crested the slope towards our houses we could see a gathering of mothers, the boy in the middle of them outside his house, on display, stripped to the waist. We went over and thankfully he was fine – he had just fainted, although he'd been hit in the arm by a projectile. We atoned for our actions and the three of us were absolutely murdered by our parents and other neighbours at the time, in a 'raised by the village' type of reaction. Frankly, it was a miracle nobody had been killed.

Joyriding was a nightmare in the area back then, along with glue-sniffing, which was also a scourge. Like zombies, the sniffers would gather in the changing shed at night or on our football pitch with their brown paper bags of glue. I never became involved in either activity and had no time for anyone who did. Truth be told, I was never interested in crime. I was no altar boy but was always terrified of letting down my family, which meant that my moral compass and my upbringing probably saved me from getting into too much trouble. My mam always told us to do good things even if nobody could see us doing them. She also said not to bring the guards to the door or she would kill me, and that stuck.

◆

In the early eighties I began secondary school in the North Monastery CBS, known as the 'Mon'. The Mon is a famous all-boys school in the heart of the northside of Cork City. It is well known for hurling and for some of its past pupils, including former Taoiseach and hurler Jack Lynch, guitarist Rory Gallagher and John Philip Holland, the founder of the modern submarine. The Mon was a place of some 1,500 boys run by the Christian Brothers. To say it was the epitome of Darwin's principle of survival of the fittest and an education in how to survive was an understatement. Punishment was swift and brutal from both the brothers and boys alike and life in the yard and between classes was like a migration of wildebeest trying to avoid the crocodiles. I was middle of the road as a student and rarely became involved in schoolboy antics except in self-defence. I did, however, tog out for the odd after-school fight, which would be arranged in the form of an 'I claim ya!' by your adversary, a challenge you could not refuse. Some fights you won, some you lost, to no major change either way in your standing – but if you didn't turn up after being 'claimed', you were fucked.

This was an early lesson on sticking to your word as a mark of honour and committing to your promise, even if you had to fight some gorilla from sixth year. Although the fight would always have a large audience, there was an unspoken code of fair play in that if you were getting battered by your opponent the mob would break up the fight. Sometimes losing well was almost as good as winning if you showed spirit and just gave your all.

My handwriting was pretty messy, as it is to this day, so much so that I write in all capitals. I had one particular Irish teacher who, being a Gaelic scholar and very neat, took it as a personal slight that my handwriting was poor. He made my life hell and would constantly give me hundreds of lines to remedy the problem, along with the odd slap. I was doing these lines one evening when my mam spotted what I was doing and went ballistic. She was left-handed and as a girl was beaten by a nun to use her right hand to write with and had experienced some of the same treatment I was getting, only worse. I guess this brought it all back to her, so the following day she turned up at the school and that was the end of the lines and the grief for me.

◆

In my last year of primary school, when I was 11, I had started to train in a kung fu school run by my brother-in-law Mike. Some of my sisters were now married, so there was a bit more room in the house. My older brother Dan was also involved and it was something I did up until my early twenties. It was my first exposure to some of the fundamentals I would later develop. The physical and combat side – but more importantly, Mike – taught me a lot more than how to fight. I learnt how to meditate and focus my energy into the physical and spiritual, I learnt humility in its most positive form, and I learnt how to be disciplined. I would train five nights a week for hours on end and learn how to use my body and push it beyond its limits of flexibility, power and stamina. This was my first major step in the journey to developing a mind-set of what is important and to shut out anything else.

With the discipline, training and lessons I was learning at the kung fu school, I decided to look beyond what the Christian Brothers could offer me. I tormented my older brother Jim with questions about the army and made a deal with my mam that I would stay in school and do my Leaving Cert and join the army as soon as possible after. But in the meantime there must be more for me out there, I'd decided with the all the wisdom of my 15 years.

And so my good friend Denis and I decided we would go up to Collins Barracks in Cork and enquire about joining the Reserve Defence Forces, known as the FCA (*Fórsa Cosanta Áitiúil*). Of course, I was under-age at the time but armed with a dodgy baptismal cert I was sure I would be able to manage it. I remember signing on and having to give my date of birth a few times until I got it right – the regular army corporal doing the swearing in had seen it all before and coached me through. He also told me to go and get a haircut and come back up the following week. And so, in the late summer of 1987, just shy of my sixteenth birthday,

I technically began my army career in A company of the 23rd Infantry Battalion FCA.

I didn't get a dress uniform right away but was issued with a set of green army overalls, which I hoped fitted the original owner better than they fitted me. They were so snug that when you raised your arms forward, your legs would automatically follow suit, which helped when learning to march.

You had to turn up, or parade, for a few months to show you were serious before you would be issued with a proper No.1 dress uniform – in that respect the FCA was more like a club than the army proper. When we were finally trusted with our No.1 uniform I was delighted to bring it home to my dad who gave me a masterclass in how to iron it correctly – not forgetting, of course, the Kiwi Polish for the boots.

So off I set in my No.1s on that first Thursday night to Collins Barracks, a good 40-minute walk away. When I look back on it, in the eighties on Cork's northside, walking in my FCA uniform amongst the joyriders and glue sniffers every week, thinking I was the business – it's a minor miracle I didn't get jumped, really.

With my No.1 uniform saved for parades and overalls for everything else, it was hard to look the part in the field. I resembled a farm labourer more than a soldier in the overalls. Although I lacked a field combat uniform I had, of course, access to my brother Jim's kit, which I would 'borrow' as required. As part-timers, the FCA clothing and kit issue we received was pretty spartan, even if it was known as the free-clothes association. Jim, on the other hand, as a regular army soldier, had some lovely gear and plenty of excellent kit.

We would parade every Thursday night for three hours or so and also did a full monthly training day on a Sunday. Every quarter we would go on a weekend training camp in the Kilworth training area or Fort Davis in Cork harbour. Then of course there was the annual training camp, which was a week away and for which we got paid a full wage. This was also when I was introduced to having a few pints in the mess, or army pub.

We learnt to shoot with Lee Enfield 303 rifles, which, although they dated from the 1950s, were an excellent weapon for beginners. The kick from the brass butt plate of the rifle was huge when I finally squeezed off the first round. Our instructors, Eddie and Pat, taught me the principles of shooting that would stay with me all my life, regardless of the weapon used: 1) natural alignment, 2) breathing, 3) sight picture and squeeze, 4) shot release and follow through. Get this right and you will never miss. In fact, in later years, during combat shooting, I learnt that sometimes two of the four principles are enough. Pat also introduced me to navigation training and reinforced the importance of physical fitness as a base for all other skills. (Physical training is a gift for life and something I still go to when I need to clear my head.) Eddie was more a military skills and procedures man and instilled the technical side of being a soldier, including deportment and attitude.

◆

I left school and completed my Leaving Certificate in June of 1988. I didn't apply myself as much as I could have and received pretty average results. I continued to parade and be active in the FCA, which I loved, making some lifelong friends there, and learning the important skills of being a soldier. I worked on a building site for a time from 1988 which was tough, but the pay was pretty good. In 1989 I got a job as a youth worker, working in various youth clubs and assisting all over Cork City with Ógra Chorcaí, a local youth organisation.

———◆———

In autumn 1989 when I was 18 years old (I'd left school the previous June) I received a letter from the army to state that I should present myself for an interview in Collins Barracks at the time and date stated. The day of my interview arrived and I decided that instead of wearing a suit that I would go to it in my FCA uniform. I don't remember much about the interview, but I thought it went well. I made sure to talk about my dad and my brother and kept mentioning the 4th Battalion, since both interviewers wore its flash (military insignia) on their shoulders.

The interview was followed by a medical and fitness test and the following week I received a letter stating that I had been successful in my application to the army and that I was to report to the command training depot (CTD) in Collins Barracks, Cork, to commence training with the 78th Recruit Platoon on 31 January, 1990. The news was bittersweet – of the five of us friends who went for interview, only two of us got in with just 40 places given from 4,000 hopefuls. And for me, this was it. I was on my way to an army career, and a continuation of the family business.

I was lucky in my early career, but my friend Denis was not. Denis, with whom I'd joined the FCA, never made it into the army and chose instead to emigrate to England to work, even though he was a born soldier. It changed his path completely in life and I met him some years later over Christmas for a beer in 1992, by which time he was a different man. He took his own life in 1994 and my last interaction with him was at his funeral in Cork.

———◆———

The things we experience in childhood form the foundations of our factory setting. I was lucky in that my childhood was balanced, stable and full of happiness and this set the foundations for everything else that was to come in my life. I learnt about loyalty and the absolute need for humour when it's appropriate (and sometimes when it's not). Being part of a large family confirmed the importance of consideration of others and that you can't survive alone for long.

From a very early age, also, I learnt that physical fitness kept my body healthy. Kung fu, running and swimming helped me learn confidence. Most importantly, during this time of my life, both during the FCA and after, I came to understand that discipline and respect are a key component of any mind-set. People should always be treated with respect, whether they deserve it or not. Discipline will keep you out of trouble – and it's not just about a toe-the-line or order-following approach: it runs much deeper. It means having the discipline of doing the boring stuff or that thankless job, even when nobody is looking.

I also learnt that throwing bullets in the fire was not one of my best ideas. Nor, in fact, was hanging off the back of a bread van.

CHAPTER 2

TRANSITION

On my first day of army training, the last day of January 1990, I woke up early. I was alive with excitement and nerves. My mam gave me a big hug as I went to pick up my gear bag before heading off. My dad told me to 'keep your mouth shut and listen' – accurate advice for life, and not just army life, I would discover.

I arrived at Collins Barracks and walked towards the gate policeman's hut just outside the red and white vehicle barrier. I had stepped inside the gate hundreds of times before in the FCA, but today was different. A friendly corporal checked my name off then told me to follow him to a building nearby. I went into a room and pledged to uphold the security of the state and follow all lawful orders in its defence and signed on the dotted line. Known in the army as your 'date of attestation', it is also the moment that the army owns you. When I stepped out of the room after my swearing in the same corporal with the clipboard was waiting. But he was no longer polite. In fact he nearly bit my face off, shouting at me to get my arse out the door and 'fall in' outside. He doubled (ran) me around to an accommodation block. If this was all I had to worry about I was laughing, I remember thinking to myself. How wrong I was.

Over the weeks that followed, I learnt the corporals were ruthless. They would jump on the slightest thing in an effort to instil absolute discipline and this involved constant, repetitive training, like marching round and round the parade square, an activity known as 'square bashing'. This is a carefully repetitive process designed to form a group of people into one unit. It teaches the individual soldier that they are part of a team and must react immediately, along with the remainder of the group, to execute the drill movement. This discipline from marching is perfected and transferred to the field and in combat. There were also lectures, weapons drill and physical training every morning and the endless inspections of ourselves, our uniforms and our rooms. We learnt very quickly that you are only as good as the weakest man, and would suffer as a platoon for the failings of individuals – a harsh lesson that builds great team spirit and group responsibility.

As we gelled as a group we found we were able to manage the training staff much better. We wore a working dress uniform for our daily routine, which had to be creased to perfection, so we used a paper-glue stick on the inside of the trousers and then pressed them, forming a razor-like crease – this worked a treat until it was time to take the trousers off, at which point the glue would take all the hair off the front of your legs as well. Room inspection was particularly nerve-wracking: the instructors would come in wearing white gloves and feel for dust in our lockers, under pipes and in all the unusual places as we stood motionless at our beds. One instructor would put guys into their metal lockers and push the locker down a flight of stairs if the standard of cleanliness wasn't up to scratch. If we failed an inspection we would not be allowed home on a weekend pass. Overall, however, although the training was intense, I grew to love the structure and the routine in that everything was mapped out for me and all I had to think about was learning how to be a soldier.

I was so close to the barracks that you could see it from the end of my road, so on my weekend pass I would drop my laundry to my mam, prep my kit as required and have a home-cooked dinner on a Saturday evening. Sunday would be spent working on my kit. It taught me that the effort you put into what seemed like a mundane and thankless exercise will set the conditions for success and have your kit in such a condition that it works when you need it to. 'Personal admin' in army-speak is not the use of paperwork or clerical activity, it is to administer or have your dress and equipment in good order and help you be prepared to operate effectively at short notice. With no women in the platoon (they trained separately then), we all dressed exactly the same down to the underpants and had all the same kit in the same place at the same time. This created the idea that we were all a part of one team, but it had the more important role of preparing us for life in the field where following discipline can make the difference between success and failure – or life or death. It also fosters attention to detail – we would stand in front of each other and correct the tiny flaws prior to inspection.

We were punished for any failings, and in some cases we would discipline our own by making them work harder for the common goal. One guy in our platoon knew all the angles and was constantly using the sick card and bluffing his way along. He never helped anyone else so was known as a *Mé Féiner*, or 'selfish' in Irish Army-speak. I always thought of my friends who didn't get into the army and whose place this guy had taken, which I found hard to stomach.

—◆—

While our sergeant carried out the daily inspections, the weekly ones were taken by the platoon commander, a young lieutenant, and on occasion by the commanding officer (CO) of the training depot. This was a do-or-die inspection, which would mean severe retribution or glory – depending on the result – so we would prepare diligently for it. 'Victory loves preparation,' as we say.

On one occasion we even had a General Officer Commanding (GOC) inspection. Now this to us was like being inspected by God himself and our training staff had us warned that if he checked us on parade and found something untoward we would be executed by firing squad, or something along those lines. We stood stock still for hours with the whole brigade on the square, overcome with cramp and stiffness, as the sergeant behind us quietly scolded any movement with dulled fury. When the GOC finally came to our rank I could see him out of the corner of my eye and braced myself. *Please don't stop at me*, I thought, *please don't stop*. He passed by me, stopped and glanced up and down as I nearly shat myself. After what seemed like forever he eventually said to the CO next to him, 'Take that man's name – he stands well on parade,' and carried on. I nearly keeled over with the relief, assimilating quickly that a pat on the back is only inches away from a kick in the arse, so don't get hung up on either.

———◆———

Finally, after 13 weeks of recruit training, on 17 May 1990 our passing-out day arrived. It was just like I remembered my brother Jim's passing-out, years before. The formal ceremony was followed by a small function in the dining hall and the army stand-up buffet lunch, of course followed by a period of the families meeting the instructors in the mess. After a few hours in the mess and polite conversation the families were allowed to take their recruit away until later that night. A dinner dance was arranged and paid for by the recruits in a hotel in Blarney and was a great night and my first date with Margaret, who I had met through a friend a few weeks earlier and managed to summon up the courage to ask to the passing-out do in the hotel.

I was no longer the glorified schoolboy who'd arrived at the barracks. I now felt a sense of achievement and belonging. I had gained so much in this time: discipline, loyalty and integrity, but most importantly I learnt to be a good follower. And this was crucial, because to be a good leader you first need to be a good follower, to support the person in control and commit to achieving the mission as directed by them. You need to understand this in order to lead people, because if you can't trust someone enough to follow them, how can you expect people to trust and follow you?

———◆———

After a week's leave for us we reconvened in the training depot to carry on with what is known as our three-star training. We were now two-star privates and would need to complete a further ten weeks of training to qualify as full private soldiers.

The training was all in the field and was based more on the tactical employment of soldiering and learning how to conduct aggressive conventional operations. It would run over the summer of 1990 and we would have a bit more leeway with time off and a pass to go out during the week up until 23.59, which is the army version of midnight. Of course, the country was in a craze with Italia '90 in full swing and we were allowed to watch the matches Ireland played in. They were good times. The experience of those World Cup matches as a nation transformed how Irish people interacted with the rest of the planet and how we saw ourselves as a nation. For those two hours of each game we watched as proud young Irish soldiers, we deserted our army and became part of Jackie's Army.

I finished my final training in July that year. My 39 fellow recruits would mostly go their separate ways to various units. I, however, wasn't alone: no fewer than 17 of us would be heading to the infantry unit my family had served in.

I started in the 4th Battalion in August 1990 and turned 19 in the same month. We were the new blood in this unit and we would all go into B Company, which was a mechanised infantry company. We were given rooms and beds 50 metres from where we had lived in our recruit training. We would all have to 'live in' officially although in reality we went home every evening and weekend and soon got into a routine of mostly 0830 to 1630 unless on duty, on operation or away training.

Just days into being in the 4th we went to a training area to carry out various live-fire practices and exercises. We mounted up in some of our Panhard armoured personnel carriers (APCs) and drove out the gate and through part of Cork City. I could see out of the open driver's hatch as I sat behind him along with eight others, facing outwards in the vehicle. The engine is inside the vehicle right in the middle, the driver in front of it with seats for three on both sides and two at the rear, with a commander in a turret on top of the vehicle. The Panhards were ancient, lumbering old French-made vehicles, which were way too loud and spat fumes and exhaust everywhere.

It was very exciting, however, as I could see snapshots of city life through my viewing hatch to the side. All was going well until, at a busy junction, we drove into the rear end of the APC in front of us. What a bang of armour on armour, with us all thrown forward in the vehicle (no seat belts) with the guy on my right pinning me to the front! That hurt. We all stumbled out a bit shook, but there were no serious injuries and were lucky it wasn't a civilian vehicle we hit as we would have crushed it. The vehicles were picked up and we were loaded onto trucks and carried on to our training area. So my first big adventure in the 4th was a car crash, literally.

I lived at home with my brother Jim so we would walk to and from the barracks every day. I really got to know the man outside our house. Jim is a slave to routine – he won't mind me saying that he'd be mightily annoyed if I was not ready at the exact time of departure or suggested a change of plan. As the greenhorn, I had to slot into his system. He would share his wisdom and experience of the army with me and guide me as to the best approach on the politics of dealing with some strong characters in our workplace.

The craic was good as most fellas in the barracks are akin to comedians so the one-liners flew hot and fast, Jim being a master. I remember a deserter turned up in barracks one day after two years of absence to clear his record, having absconded from the army brass band. Jim, as quick as anything, quipped, 'He came back to face the music!'

Life is cruel and the army is worse, they said, with the slagging furious. I loved the infantry and as new guys we were put into everything going on and were expected to perform. The senior NCO of our company, CS Scott, was a company sergeant who was a tough task master and took no prisoners. He was incredibly hard on any indiscipline, but I personally loved the structure he imposed on us. We knew exactly where we stood. I signed up for everything from cross-country racing and swimming events to military pentathlon, which is a five-event Olympic-style competition involving swimming, shooting, running, an obstacle course and so on. It was a great way of training.

Later I tried out and made the battalion falling-plates competition team, an event that blended accurate shooting, speed and composure. Six men with Steyr rifles would sprint 50 metres to a firing position, load and shoot at ten steel plates a distance of 200 metres away. The object of the event is to knock the ten plates as quickly as possible and the team that does this first wins. Two teams went head to head in a best-of-three format trying to knock the 12-inch square plates first. I loved the sound of the round hitting the plates with a huge 'ping', which left you in no doubt that you had hit it, or missed if there was no sound. It was a spectator-driven event, which was very exciting to watch as you could see the plates as they fell for each team with usually a split second between winner and loser. If the six men were firing more than two rounds each, you weren't winning, and more than four rounds – well, you may as well bayonet-charge the plates.

Our coach Gerry Conroy was a legendary training sergeant and himself a former all-army marksman, which means he was the best shot in the entire army. He had a brilliant if unorthodox approach to coaching. Gerry would give us the go; as we fired he would roar 'Steady!' to instil composure for that first second; and if I blasted my first round and missed the plate he would often kick me in the arse. Eventually after a lot of practice I developed a system to keep myself composed and take that extra half a second to focus my mind prior to the first shot. Clarity of thought must come before speed of action as accurate shooting is completely focused on confidence, skill and concentration.

Gerry introduced an invaluable process called cognitive visualisation, now an important component of the mindfulness technique. It involves visualising down to the finest detail each action of an event or procedure to basically align your brain and body to carry out the action perfectly. It comes back to discipline, to constantly practise an action over and over again to perfection. The technique can also be used in a retrospective manner to go over something that has not gone well so you can visualise your mistakes and learn from them. It works for almost any task or event and is used also to increase resilience to stressful situations by actually walking through what you have to do. As a colleague told me recently, an amateur will practise something until they can do it right, but a professional will practise until they can't do it wrong.

Gerry also taught us to think of a buzzword or phrase, and say it over and over again in our head as we fired. This helps to shut out everything around you and focus completely on the action of taking that shot. I use 'sight picture – squeeze', focusing my mind and body in that moment to concentrate on the two most vital parts of the action of firing an accurate shot while under pressure. (Later, in Special Forces, I would understand this as 'forming a bubble' where, helped by muscle memory and mental focus, your body will eventually automatically adopt the position you require.)

It is possible to train and prepare your body and mind to form a physical reaction to any situation. In the military we train and drill team or individual reactions to different possibilities that might occur during an operation or mission, such as a vehicle anti-ambush drill or a reaction to a team casualty. This is known as an immediate action (IA). But the most important action is composure and control. If you have this, it will set the tone for everything that comes after.

In the winter of 1990, for example, I was a member of a guard platoon that would be completing a period of providing security in Portlaoise prison. We guarded the prisoners, who were mainly IRA and INLA terrorists, and would often have to escort prisoners to the nearby hospital for treatment. One day about to go on a patrol, as I was standing outside the main prison gate, a woman approached. As she passed, she gave me a withering look and from two feet away spat directly into my face while she called me a 'Free State bastard.' I managed to say thank you in my best Cork accent as I wiped the spit from the side of my face. My composure and control bought me time to react in a manner that was my only option, taking the sting out of her actions and lessening its impact.

———◆———

At the beginning of 1991, just a year or so after I joined the army, I applied to serve overseas with a United Nations peacekeeping mission in south Lebanon, which at the time was the main focus of Irish troops overseas. I was selected for the mission and would commence training in the spring of that year.

Before I started training for Lebanon I had to fulfil a promise I had made to my good friend Foley, with whom I had been in the FCA. He had just joined the United States Marine Corps and was still in his recruit training in Parris Island in South Carolina. One day, out of the blue, he phoned and asked if I would come to his graduation. I said I would and hung up the phone. *Where the fuck is Parris Island?* I thought to myself.

So along with another great friend, Trevor, we turned up at his graduation ceremony in Parris Island, both of us in uniform – me in a borrowed, full army piper's uniform of kilt and complete regalia – one of the conditions of my turning up, put in place by Foley. It was great to be there for him and to see him in his US Marine 'dress blues' uniform on his big day, even if he did make me wear a dress for it. It was a great adventure to get there, but was so worth the journey.

On my return to Ireland I formed up with my platoon in Cork and commenced training there prior to joining the rest of the company and later the 69th Infantry Battalion of UNIFIL (United Nations Interim Force In Lebanon). The training was long and part of it was conducted as a member of a special route clearance team (SRCT) searching for land mines and IEDs (improvised explosive devices).

I packed and shipped my kit, said goodbye to the family and Margaret and boarded the plane to fly to Israel and truck up and across the border into Lebanon. My destination was a buffer zone called the Israeli-controlled area (ICA), manned by compounds of Christian, Israeli-backed militia, and the site of flashpoints between these forces and the various Lebanese elements such as Amal and Hezbollah. We, from the UN, were smack in the middle of it all.

I was full of excitement. It was my first overseas mission – a six-month tour with a two-week break, when I would travel to Cyprus to meet Margaret. It was exhilarating, as a 19-year-old soldier, to land in Tel Aviv and drive up through the Holy Land and cross the border into Lebanon. I had seen my dad's pictures and heard Jim's stories. I would soon have my own, I hoped.

———◆———

The landscape of south Lebanon is not unlike that of the West of Ireland, with rugged hills and deep valleys, called wadis, strewn with rocks and boulders. The local people, mainly subsistence farmers, scratch out a living on small arable farms with herds of goats freely roaming the land tended by young herders as in biblical times. The Irish have a special relationship and kindred understanding of these people, going to great lengths to protect them and help the communities.

I arrived in a village called Haddatha, the main position of our six company locations and also our HQ. I would be posted on all six locations in the tour for around four weeks each, starting on a checkpoint to the north of the village. We were put up in a fortified house nearby and spent our days and nights searching Lebanese vehicles for weapons and explosives. I was learning a little Arabic and interacting with the locals who passed through our position, which dominated three main roads at a junction.

A checkpoint at a road junction in south Lebanon is a snapshot of the local social structure. Right off I saw that a lot of men sat around talking and smoking cigarettes while the women seemed to be doing all the heavy lifting – a woman walked by us one morning on the way to market with a 40-kilo bag of flour comfortably perched on her head. I also saw a car driven by a man with five children sitting in the passenger seat and the back seat full of sacks of grain. The wife was sitting in the boot with her legs hanging out and waved at us as she went by. The young men, aggressive and keen to show off, would drive really fast towards the checkpoint in older BMWs and Benzes. When we pointed machine guns at them they usually slowed down and stopped for a search, just showing off like young lads everywhere. Every evening, just before sunset, all the eligible young women and girls would parade up and down the village for suitors to check out – no interaction though, as mixing of the sexes was absolutely forbidden until marriage. The first weeks were uneventful and, other than the odd sound of distant gunfire now and then, there was nothing to report.

————◆————

You'll always remember being shot at for the first time. The sound of that crack and thump of the bullet as it passes close by – or strikes near you – is unnerving. Explosive, artillery and mortar fire has a similar, more devastating effect of course, but I always found incoming small-arms fire more intimate and personal, I suppose because the shooter can usually see the target.

The ICA is a swathe of land that runs across the very south of Lebanon from east to west and is up to 20 km deep in parts. It was created by Israel to prevent attacks to their settlements in the north of the country, from Lebanese groups like Hezbollah and Amal. The line between the ICA and Lebanon itself is protected by a long line of Christian militia-controlled fortified compounds and checkpoints. These compounds stretch from the Golan Heights and Syria in the east to Naqoura on the western coast, snaking a line across the south of the country. The compounds are a kilometre or two apart and dominate the high ground as they look down on the Muslim villages in the wadis below.

The first incident of note took place a few weeks into the mission when Hezbollah fired at the Israeli-backed Christian militia compound nearest to our post, about a kilometre away from us, Hezbollah using the cover of the nearby village. This resulted in the Christian militia returning heavy fire from the compound and indiscriminate heavy-machine-gun fire into the village Hezbollah had used for cover. The Muslim villages were relatively sparsely populated but still had plenty of families and of course children in them.

Red tracer bullets were bouncing down the road as we dove for cover. Once I got my head around it I was grand, and we continued to report and then carry out our own immediate actions for defence and observation. We were in a protected position observing the action when the militia rolled out their T-55 tank. The sound of the tank round being fired is absolutely earth-shattering and will leave you in no doubt as to what's happening. As civilians from the village poured to our location for safety, we accommodated them in our bunker and post, which, though small, protected quite a number of villagers.

But a woman and child were trapped in a house in one of the targeted buildings. One of our corporals, Ger, went to get these people. Two of us covered him as far as we could until he disappeared out of sight behind walls and buildings amidst the crack of machine-gun fire and thud of tank rounds. We couldn't see him for what seemed like an age until he appeared, dodging back down along the road with the woman and child in tow. They were unhurt and we took them into the bunker for safety. I just looked at Ger in awe. It was an incredible act of bravery and self-sacrifice, which inspired me for the rest of the tour. I never heard him mention it and of course officially, nothing was mentioned either. He'd simply used his own quick thinking and immediate action to deal with the situation.

◆

One of my tasks was to carry out regular route clearance, or mine sweep, from our HQ to a outlaying position known as Hill 880, where we had a small post. The daily operation, called an 'early bird', consisted of a four-man team commanded by an officer and supported by an APC. It was a methodical and deliberate task that required good discipline and diligence. Getting it wrong, or taking short cuts, could mean an instant death – as had happened to the Irish previously when a young officer had been killed by a booby trap on the same route.

It was too dangerous to follow the off-road tracks used by the local goat herders or olive farmers, so we kept to the new, hard route each time, checking for trip wires as we went. (Booby traps and IEDs would always be placed on tracks, so where possible we would avoid them and create new routes we knew were safe.)

One day the officer with us was trying to rush the sweep, which you just don't do. As we came onto a part of the road with a hard right-hand turn on the dirt track, the point man gave the hand signal to halt. I went up to him and he gestured to a large flat slab of rock, which was, ironically, the shape of a small grave stone, lying in the middle of the road. No Lebanese civilians used this route and, it being summer, the ground was hard so covert digging was not really an option for anyone planting a device. The stone was new and suspect, so we marked it for our engineer team to blow up and moved back. Meanwhile the officer, not wanting to wait for the engineer team, marched up to the slab. I could see what this man was about to do and the corporal shouted for him to stop. Ignoring him, the officer walked around the slab, then crouched over it, got his two hands under the lip and proceeded to prise it up and peek under it.

The three of us left behind flattened out on the ground and I covered my head with my hands as I waited for the explosion coming any second, but it didn't. When I looked up the officer had manhandled the slab off to the side of the road and was pushing it over the edge of the wadi running alongside the road. It turned out that there was nothing there, but we didn't know that initially and we had to restrain the corporal from attacking the officer. The incident showed how lazy some people are about discipline and how sometimes the person in command doesn't always know what they are doing.

—◆—

Throughout the long trip, letter-writing was the main form of contact with family. One day, however, I managed to book a radio phone call home to my mam – imagine! Actually talking to her! It may seem like the Stone Age compared to now but with no mobile phones, no proper Lebanese phone system and the military all using radios, it was a big deal. I booked a ten-minute time slot in the MP HQ where the system, called the Rear Link, was located, with a weeks-long waiting list to get on it. The phone system was an operator-assisted call in Ireland made from the radio phone in Lebanon.

The system is antiquated and involves you having to say 'over' each time you finish speaking, like having a walkie-talkie. This is done so the operator can change over to the other person in a transmit-and-receive manner. I was delighted with myself on the Saturday morning I made the call and after some connection issues I finally heard my mam on the other end. She was so excited and emotional to hear my voice but the ten-minute slot was mostly spent with me saying, 'Hello Mam, you have to say "over" when you finish speaking. Over.' She would start talking and I'd hear nothing as she just couldn't get the hang of finishing each sentence with an 'over'! It was like a Monty Python sketch, but also really sad as time ran out so quickly. I left the phone booth feeling pretty disheartened, never braved it again and just bought more writing paper.

A strange and sinister event occurred on a checkpoint in a place called Beit-Yahoun, the last UN position along the main road and route before you enter the ICA. Myself and a pal named Paul were manning the road in the late summer afternoon. It had been a quiet day, with little or no traffic on what we called the oven shift from noon to 1800 in the summer heat. We were standing about two metres apart, chatting away, when suddenly we heard the crack and thump of a single .50 calibre bullet whizzing between us. Immediately we dropped to the ground and crawled behind a wall for cover as we waited for more incoming rounds. A bullet like that would rip you apart if it hit you.

The round had impacted on a protective T-wall (or Bremer wall, as it's also known), a 6-metre high, 2-metre wide reinforced portable concrete and steel blast wall protecting the checkpoint from gunfire. The bullet struck the wall just behind us, taking a lump out of the reinforced concrete. We lined up the trajectory of the round from the impact site and concluded that the shot had come from the Christian militia compound 700 metres away. It was obviously some militia guy who, bored witless, had decided to take a pot shot at the two Irish boys down the road. But which one of us was he aiming at? A reminder again that when you forget where you are and all is quiet the world can erupt very quickly. We stayed down for a few minutes and then tentatively got back to our feet and returned to our checkpoint duty. This time we stood in a position where we were protected by the T-wall, in case your man got bored again.

———◆———

Towards the end of our tour my discipline began to flag, which was, after six months on duty, inevitable. One day a pal of mine, Dinny, decided to teach me how to drive: the problem was that the only vehicle we had at our disposal was a APC – perhaps not the best car for a learner driver to begin with.

All was going well until I hit the gas instead of the brake and succeeded in driving the Panhard straight into a blast wall. From my open hatch in the APC I watched helplessly as the blast wall teetered and rocked, as if in an earthquake, and eventually came down with a crash that was as loud as an explosion, thankfully just missing our dining room. The senior corporal, who was the proper driver, was livid, calling me every name under the sun and shouting and yelling. It was already the second APC crash in my short career.

I was convinced I was in the shit, as a charge – or disciplinary black mark – on my record from overseas was a serious offence. What followed was the greatest cover-up in Irish military history. A plan was immediately hatched and carried out with the utmost precision and speed as the platoon acted as one to tidy up the situation. The Panhard was repaired and the T-wall repositioned with lads getting stuck in everywhere. It was unbelievable how all the boys jumped in for me and, to a man, kept the story from management. It has never been discovered, by the way, and remains an amazing example of team effort and a calm and measured reaction to something over which others might have lost their rag.

And so I came to the end of my tour, as the platoon boarded the Aer Lingus big green bird that would fly us to Dublin and on home. I arrived back in Cork by train, taking in from the window my beloved view of the northside where I grew up, and on to meet my family at the station. Dad told me that my mam nearly burned the house down with all the candles she had lit during the tour for my safe return.

———◆———

Army life and its daily routines and exercises went by very quickly. I was aware of the ARW and had seen some of them from time to time in their green berets in the Curragh camp. It came face to face for me in the autumn of 1992 and they stopped by with a roadshow to drum up students for the upcoming selection course. The ARW brief was delivered by two NCOs and a captain with a beard like a biker: they spoke of Special Operations training, hostage rescue, parachuting and combat diving, sniping and all kinds of high-speed stuff. The operators all looked amazing with fancy camouflage combat uniforms, beards and the coveted green beret on their heads, the symbol of Special Forces in Ireland. These fellas are like rock stars, I thought to myself. It all sounded very doable and exciting and by the end of the brief I was hooked, so much so that I decided to apply for the selection course. Although

I had done no specific training I believed that I was fit enough to pass if I put my mind to it and spent the next month before the course getting as fit as I could.

On the day in question we packed our kit and were dropped to the ARW HQ for selection course 'Zulu' (each course has a designation of either a letter or a number that varies from year to year), which would begin at 1800 on a Sunday evening in October 1992. The one good tip I got from an ex-Ranger was not to bring anything with me that I couldn't carry in my backpack on the first night.

That night well over a hundred of us waited in the car park outside the ARW HQ in the Midlands. There were men from all over the forces of all ranks, some like me with just what they could carry on their backs, others carrying suitcases as well as their packs. I even saw a guy with a full 90-foot climbing rope tied onto the top of his pack. I wondered what he knew that I didn't.

At 1800 on the button the gates of the compound opened up slowly like the black gates of Mordor as a small column of instructors marched out and formed a single file. They just stared at us menacingly. A roll call followed, and then came a period of complete chaos as each man's kit was checked by several directing staff (DS) and kicked all over the large room we moved to, until there were mountains of kit all over the floor, mixed all over. Meanwhile recruits were ordered to crawl or do endless push-ups and burpee jumps, as an attack dog somewhere bayed and barked.

We were paired up by a DS calling out numbers from a list after which you had to find your buddy by number, and with 100+ others doing the same thing, it was intense. I could see that some had lost control already and ran around aimlessly shouting for their buddy, while others quit there and then. We were brought out into the night in groups and made to run for hours on end in an exercise called 'scratch'.

Scratch is a physical event designed to get rid of the participants who won't last. It involves running, carrying logs and sandbags, carrying each other and crawling through mud and water. You don't know when it will end or what is next. It's hard on the mind as well as the body.

We began by carrying everything we'd brought with us on our backs and running over the plains of Kildare. We then went to a location for testing off-road military vehicles where we crawled through deep water and mud. All the time men were quitting under the pressure of this, while DS staff shouted in your face constantly. All I said to myself was to keep going as we picked up logs in teams and raced each other for hours, transitioning from exercise to exercise without a rest. The numbers were fewer and fewer and eventually the survivors were brought back into the large hall where we were made to stand in lines. As we tried to stay focused, suddenly large spotlights were switched on in our faces as a voice dramatically welcomed us to selection. *What have I let myself in for here … and what's coming next?* I asked myself. I could see the steam rise from the body heat of the students in front of me.

We were brought out again, hosed down with a fire hose to remove the mud we were now caked in and given hot food, then led into three large rooms to sleep on old army mattresses, complete with questionable stains. There was no sleep as we had to prepare and clean our kit for the next day, which came much too quickly.

In the morning we were driven to the mountains, not even able to sit for long as a DS on the truck would make us stand at intervals as we drove so we were denied comfort in any form.

The next part of the process involved us arriving in the Wicklow mountains to run what is known as 'table mountain test'. Nowadays this test involves completing a certain route within a certain time while carrying a rifle and 45lbs in a main pack. Back then we had to carry all of our kit and I personally had to carry a general purpose machine gun (GPMG) that weighed 24lbs by itself, never mind the other equipment. US Special Forces guys were there as observers, just to add to the pressure as the DS would put the boot in us. I struggled with the extra weight but two other guys gave me a hand and took turns carrying the machine gun with me. We made it back down after an hour and a half and were told it was a shite effort and to stand by to go again, which we did. Some quit there and then but the majority gave it their all again hoping it was a test and we would be called back soon. We weren't, of course, and had to go all the way up and down the mountain again. It took every ounce of my soul but we finished.

Another test involved a poncho-raft lake crossing. We buddied up and had to make a raft by tying our two main packs together using para cord (string) and branches to float it across the freezing water. We would then wrap our two ponchos (a large, waterproof cover) around the packs and tie them, creating a raft. We had to carry the heavy raft to the water's edge and wade into the freezing lake, some naked and others in shorts. To add to the madness we suddenly spotted a crocodile head on the surface of the water, which scared the life out of a few who hadn't copped on that it was, in fact, a stuffed head with a diver under it! But we were so tired and cold that it took our pressured brains some minutes to realise that crocodiles aren't usually found in Wicklow. Struggling back to the bank and into a forest to dress, my kit was soaked as our raft was crap and half sank. I was in a bad way and on the point of hypothermia when a quick-witted DS spotted me. He brought me onto a track and ran me up and down while feeding me hot tea each time I got to him, until I recovered sufficiently to carry on. His intervention and quick handling saved me.

The lessons were harsh and well learnt. At night the DS would give different classes on a multitude of subjects. I remember being squeezed into a lecture room, sat on the floor with the lights switched off – which meant, of course the battle to stay awake would commence – watching a 1970s British Army training film about hypothermia and cold injuries. I could have done with it a bit sooner, I thought to myself as a young David Jason from *Only Fools and Horses* put his frostbitten feet up some fella's jumper. *How can anyone take that seriously*, I thought. *It's Del Boy, for fuck's sake!*

In my struggle to keep my eyes open I must have drawn some attention to myself because next thing I felt someone grab the back of my neck. I bolted upright with the shock of it to find a DS standing over me in the dark, putting something into my hand and cupping my fingers around a cylinder about the size of a drinks can. I could feel some sort of a ring on top and realised to my horror that he'd handed me some kind of grenade. The DS proceeded to pull the pin which armed the grenade, ensuring that my holding the flyoff lever was all that was stopping this thing from going off. That got my focus and attention. I spent the rest of the film clutching the grenade, terrified I'd drop it. It was an efficient way to focus a sleepy student, and when the lights went on he simply reinserted the pin in the stun grenade and took it back.

———◆———

The course rolled on for a week and I was pretty ragged by the weekend when it came to a break and we could rest for 24 hours or so. They fell us in, then basically surprised us by telling us to get out of the compound and be back the next evening to carry on with the activities. I got the train with some others but would have walked to Cork if I'd had to. We burned off so much time because we had to do the same journey back, which would mean precious little time at home. On my return to the Ranger compound after a night at home I'd struggled to motivate myself – the break being, of course, a test to see how we would cope after a comfy night away. Sometimes a break can be harder than the course itself. I had also picked up a knee injury on the previous phase so I was starting to feel quite sorry for myself and was questioning my resolve.

One night we carried out an evolution. An evolution, in army terms, is a process of being put through a class, exercise or training event: this one involved instruction on command, how to lead patrols and so on. I was completely out of my depth, terrified I'd be asked to be in charge, and in my head I began to worry about the what ifs. What was next for me? The voices in my head were screaming. Why I was doing this? Who did I think I was? This is

Special Forces. I'm kidding myself here. Am I able for it? I limped through the rest of the week, knowing I hadn't the heart to keep going.

DS staff always quickly segregated those who wanted to quit from the main herd because it does neither any good, so next morning I approached a DS and told him I wanted to quit, or return to unit (RTU), as it's known. He pulled me aside and said he would give me five seconds to reconsider, but I was done, although I appreciated him giving me the chance to reconsider. I didn't quit because I was injured, I quit because I wasn't good enough.

I was taken to the course officer for an interview, where the DS told me that he was surprised and disappointed to see me go and that I was doing well in their eyes. At least my composure was good and I was projecting the right vibe – even though on the inside I was hanging out my arse. I was directed into a small hut to await pick up and just got into my bag and slept as much as I could. Another DS accompanied me to the dining hall for a meal and while I was there, in came the remaining students for their dinner. They all sat silently on the far side of the dining hall from me, I just wanted the ground to swallow me up there and then.

Later on, lying on my bunk waiting for pick-up, a DS called Danny came in and sat next to me. He told me I was too tired to feel anything except for craving rest and food. That would last for a little, he said. When you start to feel a bit more human you'll understand what you've just been through. And when that happens you will either blame the ARW and turn away for good or come back and try again. He was right, and I never forgot what he said that day.

I took some leave, got my knee sorted and spent the time going over what happened. In the end, I decided that I had failed the course for the following reasons:

- I failed to prepare properly for the course and didn't give myself a fighting chance. Confidence is built when you prepare yourself and your kit, as you gain skills and mental and emotional strength from that.
- I failed to fully commit to the course. My attitude was that I would try it out, whereas I should have been thinking that I'm going to pass this no matter what!
- Selection showed me that I had too little faith and self-belief in myself to succeed as a leader, which you only really understand when you get to the bottom of the well. I was worrying about things that were completely outside of my control, thus undermining what little confidence I had.

All my career so far I was an achiever, I was fit, I was a good soldier and had never really been outside my comfort zone, therefore I wasn't able to deal with failure.

There are times in our life when we arrive at a crossroads. In the military this is called a 'decision point' or a place where you have to know what you are going to do by the time you get there. For a long time I couldn't even think about the ARW but my friends in my battalion set me up by slagging me about the course, which is the army reaction to most trauma. I eventually stopped feeling sorry for myself and got back on the horse and licked my wounds for a bit and decided that I was going to do this again, but this time I would correct the faults that set me up to fail.

I knew I was going to bounce back from this, but it was down to me as to how high I would bounce. I realised that confidence would come from experience, so I targeted some events in my career to get this, such as more overseas service. The leadership I could fix with an NCOs course and getting some leadership experience under my belt. I knew I had to learn how to plan effectively and how to carry it out. The most important thing to me was that I had the will to succeed. Now I was on a mission.

CHAPTER 3

SET THE CONDITIONS

was now looking down the road a little and thinking about what I needed to do, working back to successfully pass selection in the future. I had a lot of distractions at home and wanted to get myself together and clear my head about all the things I wanted to do, as the failure on Zulu had cut me much deeper than I thought. My relationship with Margaret was serious and we were moving forward with our lives. My dad had been extremely unwell and just recovered from a heart attack so he was always on my mind and of course I was thinking of my mam worrying about me if I went away again.

But I pushed my concerns to the back of my mind and deployed overseas again, leaving in spring 1993 to return to Lebanon. It would give me the opportunity to train physically and some more operational experience and head space.

Some of my good friends travelled also and of course I made some new ones on my deployment in C Company of 73rd Battalion, based in Bra-shite (real name). It proved to be a busy deployment and we had our hands full for a lot of the mission with plenty of artillery shelling from the Israelis.

The communications had improved since my first tour and I was now able to pay for calls home on a satellite phone. I still looked forward to letters as you could read them over and over. My dad would send me Thursday's *Cork Evening Echo* newspaper, which contained a supplement called 'The Downtown' showing people out and about the previous weekend, so we could see who we knew. The package was folded and wrapped to perfection and covered in brown paper and inside would also be a handful of Barry's tea bags in a little plastic bag. He understood exactly what small things mean when you are overseas and away from home.

Three of us were resting in a small Portacabin one morning after a night duty, in an isolated post positioned on the main infiltration route for terrorist elements – a pretty hairy place at the best of times, designated as UN post 6-20 but known to us as 'the black hole' – when we were woken by the sound and impact of incoming rocket and machine-gun fire.

Just as I clambered out of the bed there was a heavy thud on the outside wall. Instinctively I sprinted to my defensive position, carrying my body armour and weapon, which I put on over my shorts and T-shirt when I reached my post. The firing was intense for a few minutes, but it quietened to the occasional burst and eventually petered out to nothing. A militia patrol directly in front of our post was being fired on, but our post was the bullet stop for most of the fire.

The post commander gave the stand down and we all filtered back to whatever we were doing prior to the contact – sleeping, in my case. As I stepped into our cabin with one of my roommates, Peter, something caught my eye. It lay in the space between the plastic wall of the cabin and the blast walls, on the sandy ground, just by the door. It was green/gold in colour, and I realised it was the unexploded warhead of an RPG7 rocket, just sitting there. It immediately dawned on me that it must have been the thud I heard as I jumped up. It was sheer luck it hadn't detonated on impact, but of course it could go at any minute. I took off to alert the rest of the post while my colleague went to secure the area.

On returning to the site I found Peter standing in the same place I'd left him, still in his shorts and flip-flops, but now holding the RPG warhead under his arm. With his other hand he handed me his camera to take a snap. 'Here Ray, get a picture,' he said.

I stayed very calm and spoke to him quietly and firmly. 'Now listen to me very carefully, Peter. Just place it back down where it was, right now, nice and slow and easy.'

'OK … OK … OK,' Peter replied as it dawned on him what he had done. He very gently put it on the ground.

I caught him by the scruff and dragged him behind the blast wall, well away from the rocket, where I launched at him and grabbed him by the throat, almost punching him in the mouth. The blast of it would have killed us instantly. These rockets are designed to take out tanks.

The engineers came and carried out a controlled explosion to blow up the RPG warhead and didn't even damage our little cabin. I thought about what happened for a long time after and felt relieved but also angry and amazed at how good people can do stupid things. Peter was in shock as the gravity of the situation took him over and he was pretty upset for a few days after but he got over it, as I did. The experience hit home again that your first reaction in any situation has to be composure and control.

We had a bar in HQ, or 'wet canteen' as it's known in the army, where you could go and have a beer from time to time, as a way to let off steam. It opened for two or three hours only and was usually great craic if you could get to it in time. It was later renamed the 155 Club after being hit by a 155mm Israeli artillery shell, which has a killing radius of over 200 metres. I was there one night for an outdoor show Brendan Grace was doing for the troops, which was great up until the point when a militia compound nearby opened fire in the area. To be fair, even though Brendan was rattled he carried on.

There was the odd punch-up there also – I was in one myself one night close to the end of the tour and ended up breaking my hand in the process. We all went home, me in a cast dodging the medics in Dublin airport with my hand up my jumper so I didn't have to be kept over in hospital. The lads ran a bit of cover for me also so I could go straight to Cork with my cast, to remind me if I didn't already know that poor discipline results in consequences. I think like everyone else I am also capable of doing stupid things from time to time when I lose focus. In both my deployments I had messed up near the end due to waning discipline and poor judgement. I needed to work on my self-discipline.

◆

On my return to Cork and garrison life I was again involved in all aspects of the infantry, which were mainly operational or regimental duties, cash-in-transit escorts (delivering cash to banks) and training in various forms. A large part of army life was also its ceremonial obligations including everything from performing honour guards for visiting heads of state to providing firing parties to fire over the graves of Old IRA members at their funerals.

I enjoyed the ceremonial side of it, I suppose, like during my father's time, when it was about being well turned out and the pomp and ceremony. I enjoyed the parades with the army brass band playing or marching to the pipes and drums of the infantry pipe band, the sound of which would stand the hair on the back of your neck. Back then on the first Friday of every month the whole brigade would form up behind the army bands on the parade square in best bib and tucker and march to the garrison chapel for an obligatory 'mass parade'.

As each company arrived at the church the regimental sergeant major (RSM) would be at the door to ensure that everyone went into the Catholic mass, regardless of religious persuasion. The chaplain, or 'padre', was just happy to have a congregation as most days he wouldn't have any clients at all. The upside for us was that he would bless and give general absolution to the whole brigade for our sins, all in one go during the mass, something that happened each month. It's an old tradition now well gone, but I always enjoyed the military band music of the parade – and of course the new soul every month was handy also.

By this stage Margaret and I were engaged and had set up home together. I went on two more overseas tours in order to get the money together to buy a better house and get married. The plan to go on selection was taking a back seat for a year or two as my life got well and truly in the way.

On my third deployment to Lebanon in winter of 1994 I was promoted to acting corporal for the duration of the six-month deployment. I enjoyed it. It's a temporary promotion for a private soldier who shows some ability and is selected by his platoon commander. I was responsible for patrols and leading a small team and this was exactly what I required to gain valuable experience.

I had a great trip until near the end when once again, I made a massive mistake.

I had been posted on a checkpoint along with two other lads in a village called Ayta-Az-Zutt, which had the main local road running through it. During a duty I decided to take the team off the checkpoint one night and stay within the small post nearby, for no reason other than comfort and laziness, as we were close to the end of our shift. I did this without the permission of any commander and when a senior ranking officer came through the checkpoint, there was nobody on the road that he could see.

It was a stupid thing to do and a complete lapse of judgement and procedure that could have resulted in serious consequences. Now for the first and only time in my army career I was paraded before my company commander 'on the mat' (on disciplinary charges). The company commander could have thrown the book at me I'm sure, but my platoon commander and sergeant stood up for me. Ultimately I would not be charged with any offence, or have anything put on my record. Instead I was busted back to private soldier.

I was disgusted with myself. I'd let myself down with the two men who had faith in me and yet again finished the tour with a bad taste in my mouth. I travelled home under a cloud, lucky not to be locked up. It was one of the lowest points in my military life as I contemplated how I had this ridiculous ability to drop my guard completely at the worst time, undoing all my hard work.

◆

I was now, in 1996, six years in the army and had really made no progression from day one. Instead, I was simply and comfortably going through the motions of soldiering. I knew I needed to get out of the comfort zone of being a private and get my ass in gear to jump-start my career, or I faced the risk of never moving on. It had been three years since I failed selection and at the time it looked like I would never get back to it. And what about my grand scheme?

It was only later that year that I managed to get the finger out and secure a place for myself on a non-commissioned officers' (NCO) or corporals' course. The five-month course took place in the CTD in Cork, for soldiers from all units to qualify them for promotion to corporal and responsibility for a team of ten. It would be the first step in learning how to become a junior leader. The course is very formal and pins the bulk of the focus on the ability to be able to train, lead and manage soldiers efficiently and effectively. It would also open the door to another side of the military, which had been a large part of why I had failed on selection previously.

The initial phases of the course are all based on instruction and the development of how to become an instructor and manage troops. The army has an age-old way of teaching you how to conduct yourself as an NCO: 'Fair, firm and friendly' is the maxim. We did week after week of training and practice on how to form a plan for a skills lesson on weapons; how to study it; and how knowledge of the subject is the key to delivering that lesson in a formal military setting. In foot drill we had to learn by heart all the individual drill lessons in Irish, projected in a formal parade manner on the square.

Along with this were periods of taking control of groups and being responsible for the management of them.

We were versed in the joys of public speaking and had to deliver test talks and briefings on various topics, which definitely increase confidence. I completely committed myself to the course, studied as much as possible on weekends and basically worked my ass off.

We were drilled on confidence: how to project yourself from a lecture to parade setting by 'booming' your voice and exaggerating your actions and on the importance of your attitude towards those you are teaching, reflected, for better or worse, by what you put out there. We also learnt about the absolute necessity of discipline, from maintaining your own to imposing the army's version of it on subordinates. Finally, we practised what is at the core of instruction: enthusiasm and manner, which basically means you got to sell it to people as best you can and believe in it.

My education in leadership was compounded by several lectures and case studies given to us by the officer in charge of the course. He introduced us to the different types of leaders, their qualities and characteristics, all discussed in great depth. This was all very much theoretical, as I still hadn't the practical experience of how to use it or a full understanding of how to employ my leadership skills on a daily basis, since knowledge comes from education and wisdom comes from experience. I did realise, however, that this would be an ongoing thing that I would develop and learn as I went along.

I watched my instructors closely and imitated some of the mannerisms they projected in both a class and management setting: the autocratic leader, the democratic approach or the laissez-faire style. We eventually moved on from all the formal training and onto the tactical side, which meant learning how to conduct operations and lead troops in the field.

I had a quick intermission in my focus as I was part of the track and field team competing in the All-Army Track and Field Championships 1996, a welcome break for two days from the course. The officer in command of my NCO course was also involved as part of the 4 x 100m sprint relay team, in which we won gold. In addition, I won the individual 100m race and became the all-army champion for 1996 – how bad.

The key part of an NCO course is learning how to plan effectively – one of the bulwarks of leadership at any level. And the first thing we had ground into us is that old maxim of the seven Ps: Proper Preparation and Planning Prevent a Piss-Poor Performance – old and accurate and so true for me.

The army also teaches a simple but effective system in planning an operation, which incorporates all the activities to be completed along with delivering the plan to your team in a formal and detailed way. The basics don't change, and it can and is used from the lower leadership level all the way to the top. The five components of this system, used by most armies, are: Situation-Mission-Execution-Support-Command/ Communication:

Situation involves all the information you have on what you are dealing with, how it affects what you are doing and your team and structures. The more information you have, the better your plan of execution will be. Attention to detail is important.

Mission is basically what do you plan to do, and is broken down into a simple format of who, what, where, when and why. A simple sequence of words that will focus your team on **Who** needs to carry it out, **What** needs to be achieved, **Where** you are going to do it, **When** it needs to be done by and **Why** it needs to be done. A basic mission-statement breakdown.

Execution. A detailed breakdown, often in phases, that projects exactly how the plan will be conducted and completed. This is normally the longest and most important part of planning and should include potential pitfalls.

Support is the detail of all the back-up required for the plan and who will supply it. The support is often arranged by the second in command (2I/C) of a team and will incorporate everything from assets, equipment, specialist tools, feeding arrangements, rest and anything else that needs to be considered to assist the plan from a housekeeping point of view.

Command/Communication is the chain of command and responsibilities of people from the top down as well as the main timings of what comes next, what backups are required for clear communication, systems, hardware and fall-backs.

This is the basis of how a plan is developed and delivered. You need to sell this with conviction. Each heading can contain as much detail as is required but keep it to the point and don't waffle. The final and critically important part of any plan is to practise or rehearse what you plan to do. This will get people's ass and brain in gear, emphasise the main points and identify potential problems.

You can apply this format for quick situation assessments or super-detailed and long-term plans, which may involve multiple people and assets over a long period of time. Assets in the military are what you have to hand like vehicles, aircraft and fire support, for example. In the civilian context it can be the same, i.e., what you have at your disposal. I have used this system in the military to plan and execute patrols and missions to deliver a plan to my team and get them to carry it out. In the civilian world I have used it to brief people in the conduct of a product roll-out campaign, for example, where we developed a plan, took a part of responsibility each and executed it to a successful conclusion. It can also be shortened to a quick brief, following the main headings and giving a one-line explanation of each part of the plan for a fast reaction or rapid alignment of your team and immediate plan. The main principles of any plan are that it must be simple, your message must be clear and the plan has to be flexible.

Oh yeah, and always include a Plan B.

Another useful piece of advice we were taught in training is plotting a timeline. You draw a line from the start point (where you are in your planning at that particular time) to where your successful mission ends, and plot both in time. In between all of this you insert all the other events and actions you need to complete to set the conditions for success. This will focus you and your team on the smaller tasks that must be achieved, and build up to the success of your mission. It gives your team a 'when' feeling, or anticipation of what will occur in the order it should occur and gives them the opportunity to prepare for the events as they arise, which will help team confidence and interaction.

It also gives them a deadline and completion date for individual tasks and goal alignment.

When we finished our tactical training we had various test appointments where we would lead attacks and patrols of our peers in various scenarios. The instructors set up intricate and testing situations that we leaders had to deal with under pressure, so creating resilience in our decision-making process.

I completed the NCO course, my core values having been, once again, adjusted. I could now understand the simple mechanism of being outside my comfort zone, having been forced to make decisions under pressure. I did well. The top three candidates were paraded separately and given awards for achievement from the general. I finished in second place overall behind a brilliant soldier, so I was happy enough to receive a plaque and a pat on the back. I had picked up a lot of experience and confidence and my ability to problem-solve and remain resilient was also improving. I was finally maturing.

◆

I was given the opportunity to instruct when the 4th Battalion began to conduct training for new recruits. I was delighted to do this as it meant that I had a chance to really practise my abilities as an instructor and teach young people from scratch. Being in the driving seat, as it were, also meant that your own standards of discipline, dress and deportment were on display at all times; and because recruits very quickly mirror what you project, it is essential to uphold those personal standards.

The platoon was made up of both men and women, which was a new challenge in how to approach certain procedures. Army regulations are extremely strict about the mixing of sexes. As a male instructor, at no time should you be alone with a female recruit. Inspections could be awkward: our female instructor wasn't always available and the kit the female recruits are issued with isn't the same as the men's, so you have to be careful exactly how you poke around the locker inspection. I was tough but fair, regardless of the recruit's gender, and would always reward effort and try to instil the values of teamwork, selflessness and spirit. The whole training process is designed to build up the recruits' skills and mind-set slowly, over a period of time, They all don't get it at the same time and as an eager and aggressive young corporal I quickly understood that it wasn't always about sinking my teeth into them. Sometimes the carrot works better than the stick and you can be kind to them in a way that shows them you are there to make them better.

I worked part time as a bouncer, where I learnt some great stuff about body language and how to read someone's state of mind from their physical comportment. I discovered the importance of tactical communications and position in order to talk to someone who is riled up and full of drink, spoiling for a fight. Relating to where people are in their head and showing some empathy for them or their situation is usually enough to disarm them. It worked 90 per cent of the time although there was always that one exception you end up putting in a hold or pinning on the ground because they are beyond reason. This lesson I also carried forward to my military experience, reading aggression during overseas missions or simply watching members of my team or trainees.

One evening I was inspecting the women's ablutions (military speak for washroom). The recruit-room orderly stood outside the door with the other five women in their dorm room. The girl was very composed; in fact, she was one of the strongest recruits in the platoon although I suspected her of being over-cocky and casual – always a red flag. She didn't give her report the effort it required and skipped through parts of it almost jokingly, assuming it was OK. They were nearing the end of their training and, in my view, slacking off (something I have also been guilty of in the past).

I stepped into the toilets – the place was like a palace, with the brass pipes from Michael Collins's days shined and polished. It was a good turnout but I needed to teach her a lesson for her casual approach and lack of discipline. She and her team had done the hard physical work but were already congratulating themselves, assuming they were done. But you can never be done.

A good sergeant once told me about the trick of always having a dirty piece of cloth in your pocket when doing an inspection, which I had. I dropped it on the floor in a corner and called in the room orderly and the remainder of the recruits and proceeded to give them a bollocking over the suspect piece of rag and how they were disrespecting myself and the army by its existence. I told them they had 30 seconds to find out who had left the rag on the floor, then I stepped out and closed the door behind me. I put my ear to the door and listened as the inquisition kicked off on the other side, erupting into accusation and chaos. I did this to show them that you need to follow procedure correctly and never to assume anything because 'assumption is the mother of all fuck-ups' as the saying goes. And to be honest, I also did it for the craic.

———◆———

In October 1996 I went overseas once again to Lebanon, this time as a fully substantiated corporal and with a platoon from another brigade. We arrived with the first rains of the winter there and to Christmas decorations and music, the parting joke of the guys we relieved. I slotted in well with my platoon – they were good lads – and had a great trip where I managed to maintain my discipline throughout. We were positioned in various outposts in the company area of operations over a wide area radiating around HQ.

This particular trip was through the winter, which sees Lebanon turn into a sea of red mud in the rains; walk anywhere and you end up with three inches of mud around your boots. I also spent another Christmas away from home, and having such a big family meant I was flat out on the letter-writing front. The winter also brought a different threat in that soft ground allowed for more IEDs to be dug in and Hezbollah were pretty active on the dark winter nights, with us getting shot up plenty of times.

I was moved back into HQ where there was a salmonella outbreak, picked up from a poorly stored egg mayonnaise, which nearly wiped out most of the company. Out of 120 troops only 30 avoided food poisoning, which meant we had to man HQ as other companies took over our outposts. It was a mad time of mass casualty activity with people collapsing and being medevaced (medically evacuated) to hospital left right and centre, some in a really bad way. I was part of the team that would stretcher these guys out of rooms and onto transport to hospital. People lost complete control of their bodily functions and I never thought I would see someone shite straight up into the air while face down on a stretcher, but unfortunately I did. That image stays with me. I'm not sure what I learnt from it, though as well as a practical introduction to crisis management, I definitely gained a better appreciation of physics and the relation of gravity to force!

I had now proven to myself that I was able to lead a team of people in a real operation and I had the basic raw materials that I needed for Special Ops. Every day on deployment is like a week of civilian life, you experience so much. I now just needed to commit to it and arrange an actual plan to go back up on selection after I got married to Margaret, which we did in the summer of 1997. It was a great time: I absolutely adored her, was very happy with my life and was comfortable with where I was in the military.

Between 1997–8 I was involved in the training of two recruit platoons and it was one of the best experiences of my army career. It allows you to share your own experiences with a group of people who are enthusiastic and eager. You have a fantastic opportunity to create a factory setting in them that will last them the whole of their career. They trust you completely: you are teacher, leader, parent and executioner all in one. It's a pleasure to watch these raw recruits grow from timid young people to a strong united platoon of soldiers over a relatively short period of time and it reinforces your understanding of how the process works from the other side. I also learnt a lot about my own abilities and how I could use them to guide and influence others as I had come full circle in my own army experience.

◆

In the autumn of 1998 I lost my dad, relatively suddenly, I suppose. A calm, gentle and deliberate man who had the most enormous influence on how I viewed the world, he died at home on his rocking chair with my mam beside him, on a Monday night in October at the age of 65. He died as he had lived, with honour, with dignity and with as little fuss as possible. It's true what they say – old soldiers don't die, they just fade away. When I arrived to the house soon after, he had been laid on the sofa in the room where he died. He looked as if he was asleep; in fact, he looked so comfortable that I undid his laces and took off his shoes.

The loss of my dad had the added effect that it galvanised my resolve to carry out my plan to pass selection, as life goes by so quickly you just have to grab it with both hands. So with no more excuses and new resolve I decided to set my goal and timeline: I would pass selection the following year. I immediately sat down and wrote out a plan with a timeline. I had a serious talk with Margaret and explained that I would be concentrating on this for the next year. She was happy to support me as she knew I had always wanted this. In fact, she was amazing. This is what I wrote:

- ► **Goal**. To pass ARW selection.
- ► **Commitment**. I will immerse myself completely in the idea and everything I do will be not just to pass but to serve as an operator in Special Operations.
- ► **Plan**. I will draw up a training plan that will see my body, mind and soul trained as best I can. I knew what had let me

down previously but I have created a strong inside voice that I can depend on when I am at rock bottom. I have gained confidence from the tasks I completed prior to the course, thus increasing my resilience and fitness, technical ability, knowledge and familiarity with my kit. I will also train outside my comfort zone and put myself under pressure as much as possible.

► **Timeline**. I give myself one complete year, broken down into monthly macros, building up to the course so I will peak at the right time.

► **Assessment**. I will add weekly and monthly assessments in physical fitness, technical skills, navigation in the mountains, leadership tasks and plans of attack. I need to give give myself checkpoints to make sure I remain on target.

There are plenty of formats on how to set and achieve goals, which I'm sure are a lot fancier than the one I wrote for myself here, but this one works. It's simple, effective and covers all angles (for any goal, whether it is a professional or personal one). The important thing is to visualise yourself completing it: don't just 'have a go', because unless you completely believe and buy into it, it won't work. For any goal to work, you have to set the conditions for success in such a way that the chances of you failing are all but nullified.

Plan in hand, I launched myself into following my goal for the rest of that year continued to use my cognitive visualisation techniques. I also kept a lid on my tendency to worry about what

I couldn't control and used the buzzword technique to turn my focus to something I could affect in a positive manner. When the brain thinks it has a solution to a problem it moves onto the next thing pretty quickly.

———◆———

In order to maximise my chances of passing selection I also applied and was accepted into an army physical training instructors' course (PTI), which ran for five to six months in the school of physical culture in the Curragh camp. It began in early 1999 and would be the final puzzle piece in how I would train and prepare for selection. There were plenty of lectures, tests and training evolutions on nutrition, goals, processes, programme prescription and all the health and fitness implications. The course is broken into various modules and has a civilian qualification running in tandem along with the formal military side of it.

A part of the PTI course I found particularly interesting was 'the dimensions of wellness', which was a technique that taught you how to connect up physical, mental and emotional well-being. It was considered a bit hippy-like by the army but to me it made perfect sense: a healthy body equals a healthy mind. I learnt that mental health is maintained by keeping your mind engaged and open to new experiences. The emotional or spiritual is simply your awareness of yourself and what works for you, knowing how you feel and why, and using that awareness to deal with life's challenges. It's the ability to channel your emotions effectively – fear, stress, frustration as well as happiness and love. We tend to focus on the more negative emotions, mainly because they have a more physiological effect on our performance, yet fear and stress can actually improve performance.

I was getting all the good stuff to strengthen myself for selection, physically, mentally and technically. We even had a week run by the ARW PTIs that dealt specifically with physical preparation for selection and how to train students for it. They put us through some of the tests and best practices for training – a bit like a selection masterclass, if you like. My focus absolute, I would train on the course all week and when I went home on a Friday would spend all day Saturday in the mountains carrying a large pack, weighted increasingly for the mountain phase of the course.

Preparing for the course taught me that physical fitness is one of the most important assets we have. I developed a simple plan, known as FITT, which is easy to adapt and change according to your circumstances:

▶ **Frequency**. How often we should train? Usually between 3–5 times a week.
▶ **Intensity**. How hard we should train? A percentage of our maximum effort, between 65 per cent and upwards to maximum.
▶ **Time**. How long we train for? A 10–15 minute warm-up and pre-stretch, 30–45-minute main event and 15-minute cool down and post stretch.
▶ **Type**. What training you do? Cardiovascular: running, swimming, cycling and walking. Strength: weight training

and circuits. Endurance: a combination of longer cardio with more emphasis on long duration. Power: weights, plyometrics, speed training … the list is endless.

Once you've tailored the FITT plan to your needs, it's time to consider the following:

- ▶ **Overload/Progression**. You need to gradually increase either time or intensity as your body will just adapt to the training. The FITT principle will help this but the rule of thumb is to increase by no more than 5 to 10 per cent per week.
- ▶ **Specificity**. The type of training needs to be specific for the goal or sport you train for. However, if it's general fitness you want, all movement is good.
- ▶ **Reversibility**. If you don't train or your intensity is not enough, you will lose what you have gained.
- ▶ **Individual Difference**. We are all different and react in different ways to different types of training, so you need to find what works for you.
- ▶ **Recovery**. Most important: you need to let your body rest between sessions. Recovery is essential to help the body heal and build after hard sessions. Without rest and recovery, you won't progress with your fitness – you'll just burn out or pick up injuries.
- ▶ **Flexibility**. This gives the body the ability to absorb impact and so become more resilient to potential injury as well as

recovering quicker. Ryan Giggs played professional football well into his forties because he did yoga and his flexibility was excellent. Always warm up and stretch and cool down and stretch before and after training, as well as giving yourself 20 or 30 minutes of separate stretching sessions once or twice a week. It will pay off, believe me. So many of us lead a sedentary lifestyle, sitting at our desks for as much as eight hours a day. But even stretching and walking around the room every 40 minutes will make a big difference to your posture, and powers of concentration. We also need to give the body back the time it needs to move, so activity for 30 to 45 minutes a day is vital. Never underestimate the positive effect of movement on our minds and body.

———◆———

I finished the PTI training mid-summer and returned to Cork as a fully qualified physical training instructor. I loved the course, and learnt so much on it. That autumn I officially applied for the next selection course for Special Forces, which was advertised as 'Golf One'. I was never going to be more ready for it.

It's normally a formality, done through my parent unit, the 4th Battalion. But I was shocked when my company commander said that he would not be recommending me for the course. He said that I owed the company for getting on the PTI course, and that my loyalty to my unit was in question. I was livid. This man was only passing through the company, ticking a box to qualify him for promotion in future; he would only spend a year or so in the position before returning to a cushier number. I had been in the company for nine years. I asked to see the battalion commander to explain my case.

The battalion commander, a man I respected greatly, followed the order of my company commander not to recommend my application for selection. I was gutted. I left his office, my mind racing. After all the training and sacrifice! As I walked down the corridor, the door to my RSM's office opened and he pulled me inside. He had been my CS for years before his promotion and always had my respect. He made me aware of a chief of staff directive, which stated that anyone who applied for selection had to be allowed to go. This directive was put in place as unit commanders would of course stop good lads from applying for selection because they would lose them. I went back to my company commander and told him the regulation and that he was obliged to observe it, which of course he didn't like. He said, 'Well I hope you pass, Corporal Goggins, because if you don't and you come back here, life for you will be tough.'

I couldn't believe his attitude. It was not typical of people I have served under before him. I was gutted. The bridge to the battalion was now on fire as I did my final prep, which I took as a source of motivation to succeed.

———◆———

My training had been perfect. I was as prepared as I could be. My technical ability I had tackled over the last weeks, working on the hands-on skills with all the equipment I would use along with my leadership, navigation, medical, communication and many other skills I knew would be tested in the coming weeks. I concentrated on my kit and clothing. I waterproofed my boots and made sure all my buttons were sewn on with fishing gut so they wouldn't fall out. I padded out my marching order (a harness system worn around the body to carry kit and equipment) so that it wouldn't cut into my hips and back on the long, weight-bearing marches ahead. Completing the small, final tasks prior to a major challenge I find to be an excellent method to instil confidence in what you are doing. It gives you a sense of achievement and well-being, and focuses the mind on the challenges to come.

———◆———

On the Thursday before the start of the course on Sunday I was home preparing for the course and getting my final rest prior to the madness. Unfortunately for me, selection wasn't the only thing I had to worry about. Some issues at home came to light that evening, which resulted in the end of our marriage. Margaret had been really supportive of me and my job and had no problems with me going off training, so I thought. I just hadn't seen it coming at all. I was in shock and I couldn't believe it. Our relationship ended there and then, in a conversation that wasn't even an argument. I asked her to stay with a friend that night to give me some space, which she did as I contemplated my life without the person I loved the most in it. For me, there was no going back. I just sat on our Laura Ashley sofa, the one we had chosen and bought together. I was crushed. I sat there in a daze, when the army kit bag in the corner of the room caught my eye, and I refocused. I had two choices as I saw it: go on selection on Sunday as planned or lie down and die.

My marriage collapsing before me was the final straw, as well as the ultimate motivator. My life as I knew it was over: the Ranger Wing was my chance to start a new one. I had been building my resilience for a few years at this stage, but never thought I would be needing it like this. It was another decision point in my life and I had no choice now but to go forward, because I couldn't go back.

CHAPTER 4

ALL IT TAKES IS EVERYTHING

After the bombshell of the events at home I became totally focused on preparation for the selection course, which was the right thing to do because without it, I may have fallen apart. My family and good friends Keller and Foley (now back from the Marines) were there to support me but what could anyone do, really? For the remaining days prior to the commencement of the course I just rested, completing maintenance training. I was completely in the zone. I got my head and ass in gear and loaded my kit in army transport to bring me to the Curragh and my chance at fulfilling my sole mission in life.

So here I was again, selection course number two, back outside the gates of ARW HQ with a load of guys who were as nervous as I was. I reminded myself of my preparation and path to this moment. I had no option, I had to pass.

The course began at 1800 on a September Sunday evening with 'scratch', the infamous, hours-long baptism of hardcore physical activity. We were all given a helmet with our number, because that's what you become: you're just a number. You no longer have a name, rank or persona and whatever you were before is wiped away. I became Ranger 22.

I remember a DS carrying out a warm-up with us on the Curragh plains prior to scratch in a spot just overlooking the newly built section of the M7 motorway. I'll never forget it because there was a tailback of match traffic heading south from the All-Ireland

Football Final that was on that day. I could see the Cork flags and banners on the lines of cars as the dejected fans went home after a defeat. I hoped this wasn't some sort of bad sign for me.

The event began with around 70 of us or so on the start line as we went into the intensity of this assault on the senses, with the pain and effort a constant. There is no complicated plan for this event, there is no 'why', just 'do', and several hours of hard physical activity. I survived the first night as we limped back to the Ranger compound for the nice fire-hose shower. This is only a leveller for the course, the real business would begin the next morning.

The process had developed somewhat from my earlier experience and although as intense, it was now much more structured, involving an average of two tests per day over the first two weeks, during which time the course thins out as candidates drop away.

During a mountain march we were supposed to carry a specific weight of equipment but one of my mates, a young officer from Cork, had put in way too much kit and was struggling badly. He was falling well behind, so I stopped and took some of his kit to help him out. We pushed on as hard as we could, but he fell further and further behind with me pushing and pulling him along. A DS eventually told me to leave him, that I had done enough and had to pass myself or risk being put off the course. I pushed as hard as I could, but I couldn't make up the ground to catch the main group, and so we both failed the evolution.

When we returned to the compound the course officer, a captain and the sergeant pulled me aside, I thought to kick me off for failing the march and to RTU me from the course. I was shitting myself as they aggressively explained that this had been an individual best-effort test, and that I should not have helped out anyone else. I relaxed slightly as I gathered, between the lines, that they in fact appreciated and respected what I had done – to risk failure and RTU to help a buddy. It's the cornerstone of what Special Operations is about. I was lucky, and remained on the course. It was also the one-year anniversary of my dad's death – I'm convinced he was looking out for me!

By the end of the first week the course had whittled down to 40 or so men from the 70-plus starters. We had a ball-breaker test, which involved a 10-mile road run in full weighted kit. I was struggling and beginning to fall back from the group on the last mile or so. A DS running beside me, an officer named Mark, took off his green beret and held it in front of me saying, 'Do you want this?' I did, badly, and it was the spur I needed to push harder and catch the group to finish with them. It was a great moment of empathy on his part to identify I needed a boost, as others would have let me fail, and confirms that we always owe so much to others who get us over the hump.

Unfortunately the young officer I helped got injured – his Achilles tendon basically tore away with a sickening, snapping sound. He was in agony and unable to carry on with the course, but he was a good fella and I was sad to see him go.

We were given a 24-hour break and I was completely prepared, unlike the last time, to manage it in a positive manner; to come back. I felt strong and didn't once even think about giving in, telling myself to remember who I was and why I was here.

———◆———

All through the next phase of the course and as the group got smaller, we began to get to know one other and gel as a unit. I met one of the best friends I'll ever have one night, in the back of a truck. A DS paired us off and I remember Staff's mad Wexford accent, and the biscuits he had, so it was a match made in heaven. We ended up in teams together for many years after, both in the military and on private contractor tasks, and we still work together on *Hell Week*. Staff is the best natural judge of character I have ever met; he constantly challenges me and has had my back more times than I care to recall.

The relationships you forge and the bond you create when you depend on someone and put your life in their hands is immense. Jeff, a good friend I knew before the course, had a water phobia. We had a test coming up where we had to jump from a high bridge into dark, deep and cold water – his worst nightmare, in other words. We managed to get ourselves in the line to jump so I was directly in front of him and would hopefully be able to help him in the water when he landed.

I jumped. They make you shout out 'RANGER!' at the top of your voice all the way down, which has the effect of emptying your lungs, so you will sink a little more on impact. When you hit the water the cold of the mountain lake penetrates your brain and the cold-water shock makes you gulp for air, as you sink for what seems like forever. The colour of the water goes from red brown near the surface to pitch black if you go beyond two or three metres. You kick as hard as you can to get to the surface and it takes a couple of seconds, which seem a lot longer, I can tell you. When you finally break the surface, you need to make eye contact with the DS on the bridge and give your name and report that you are well, all to test your composure.

Once I surfaced and had given my name I swam towards and under the bridge, pretending to be slow to wait for Jeff, who'd hit the water like a buffalo and come up in a ball of shite. He managed a report and I grabbed him and pulled him to a rope that ran parallel from the bridge to the shore 100 metres away. He was under severe pressure so I pushed him to where he could pull himself in. All the while I swam beside him because if I grabbed the line it would sink more and pull him down and he would panic even more than he already was. We managed to do this as quietly as possible so as not to draw too much attention from the rescue swimmers and DS around us who knew our game, then ran back up to the bridge do it all over again, which Jeff managed.

The next phase, the ground phase of the course, involves patrolling in the mountains for two weeks, during which we are tasked and tested both as a group and as individuals. With minimal sleep under our poncho tents, carrying out operations day and night and living on dried army rations and stream water, this was a tough call for anyone.

Being constantly wet, I had one dry thermal top, which I kept in the zip pocket of a warm fleece jacket in my pack and put on just before crawling into my sleeping bag. We would sleep under the canvas poncho, weapon on our bodies with boots on fully laced up and completely dressed with an arm through our carrying harness, so if we had to get up quickly the harness is on you with your equipment and you are ready to fight. Once forcibly woken, the hard part was taking off the lovely warm top and putting on the wet one in the dark, the cold of it against your skin.

We had so much to carry that we would cut down on weight by having just a dry and a wet set of clothes and if the dry set got wet it was torture. I would stuff my wet socks under my armpits and groin to dry when I slept. With constant wet boots and feet, the great joy of the day came if you got to take your boots off for a minute, powder your feet and put on semi-dry socks. This was done one foot at a time, with boots instantly back on. Eating was also completed as quickly as possible and equipment immediately packed away after use. There was no washing at all, just the chance to brush your teeth once a day, so we looked and smelled pretty rough.

I was dragged over a mountain for hours by another student during a patrol, an officer who was also patrol commander. We were in the Slieve Bloom Mountains, which are full of peat hags you have to scramble over and peat bog you sink into up to your neck in the darkness. The ground was hell: I constantly sank into the peat and had to be pulled out every time. I was the radio operator for this phase so had to carry extra communications equipment, which meant my pack was much heavier than most. I had to be near the student patrol commander at all times so we were buddied off, which meant he was the one who fished me out of the bog and vice versa. I don't think I would have made it without his help. We arrived in a forest sometime the next day and as we set up to rest, he just quit. I couldn't believe it – he had been an absolute rock for me and seemed to be cruising. It just goes to show that you never know how someone else is coping.

I was tasked as patrol commander and after another night march we made it to a forest, where we had to carry out drills to secure a perimeter before we could climb into our sleeping bags and pass out with exhaustion. Minutes later I was woken by the report of a single gunshot (our weapons were fitted with blank firing attachments, which allows the weapon to function as normal without firing an actual bullet). The bang came from J – also to become a lifelong friend – who'd squeezed the trigger of his weapon while asleep. Thankfully the DS didn't make us move location, which would have been a nightmare. He did, however, make J dig a grave for himself in the middle of the forest, following which I, as the commander, had to deliver a eulogy and funeral with full military honours, actually burying him enough until the DS stopped us. The lesson was effective, I might add.

We got some sleep and then started to prepare for the next mission. That night we were split into two groups and given a huge, additional supply of ammunition and equipment to enhance our march over yet another mountain. My 2I/C pulled a fast one and stowed all the resupply into my group's main packs. He should, of course, have distributed the resupply between the two groups. We were trucked to Wicklow and started walking the route that would bring us up and over Lugnaquilla, its highest point.

It was torture carrying this ridiculous weight, well over 100lbs, one step forward and two steps back, for 16 hours of absolute pain and misery. That type of slog is very hard on the soul and you go to a dark place to keep yourself going. Like any hard task I break it down to small parts: get to the next boulder, get to the next water stop. If you think about the full journey or task it will tip you over the edge.

It's a process that can be applied to any tough task, to break it down to smaller manageable micro-tasks – small bites of anything will add up and this will build your confidence as you go on.

At times of great physical stress such as these you drop below your normal range of mental and physical performance and your brain will send you messages to stop or quit. You go into autopilot and carry on, or you quit. The thought process is broken down to the simplest of actions and you develop a system inside to keep yourself going. Some think of something else, a mantra or words repeated in their heads. This is done to block out physical pain or events that may be occurring around you. For me it was a song, in particular 'Our House' by Madness, repeated over and over, which allowed me to focus on something other than the effort and pain. I concentrated on the fact that this was not going to last forever and try to focus beyond it, like thinking of the moment when I would zip up my sleeping bag and life would be good. I just focused my entire being on the mechanics of taking the next step.

Just after daylight we found ourselves patrolling across a small river two or three feet deep. I remember the effort of fighting through it in an exhausted state. Suddenly I stumbled on a rock and fell forwards into the river. The fast current pushed my pack down on me and with the extra weight, I couldn't get my head out of the water for a desperately needed breath, so I was starting to get a flap on as I struggled to roll over or even kneel up. I fought and fought for breath in the freezing river, then the next thing I knew my pack and I were both pulled up out of the water so I could get a screaming breath of air. Jeff stood there like the mountain of a man that he was and proceeded to pull me to my feet. This time he got me out of the water.

We carried on with missions and operations for some more days until one night, all sat on our packs in the pitch black waiting – waiting for what, we didn't know – the course officer called out a name – my 2I/C, as it happens – and curtly told him he was off the course. They pulled him out right there and then. They knew what he had done with the resupply and some other things and just let him carry on, knowing they would axe him. The idea of selection is to find the core values in someone that form the basis of an operator, such as integrity – all the skill and techniques can be taught later, once you have the platform – and that guy was short on it. Integrity stops us from being selfish or dishonest and pushes us to do the right thing, a good offset to being an asshole.

The following morning, we were marching again in full kit before first light, but this time it was on a proper road, so we knew it must be the 42km walk back to the ARW HQ and the last event of the course. It was a tough walk after weeks of hardship and intensity, but you can push yourself much further than you think – there's always something left in the tank. We arrived at the Curragh camp in formation after several hours of hell and of course we had to run the last couple of kilometres to the ARW HQ, because that's how we roll.

There is no sensation like crossing that line after all the degradation and effort of not just the last few weeks but also for the years I had had this one goal. All the available unit members were there to clap us into the Ranger compound, lining the entry way on each side. We ran past them with newfound strength and a feeling of pride. We had made it to the finish line. We stopped and formed a line. Then every unit member shakes your hand and you're given a cold beer. I was afraid to sit down because I knew my knees and back were finished and I wouldn't get up again (in fact my knees were never the same again and I later got laser surgery to right them).

We were all emotional in our exhausted, post-adrenaline state, and just stood around in disbelief of this being actually over. We couldn't walk properly, lads had no skin left on their feet or shoulders, our bodies were broken, we were mentally and emotionally drained, but we could stand. All it takes to pass is everything. The next day we were presented with our *Fianóglach* tabs, which are black and gold insignia worn on the left shoulder. It was a proud moment. The badge signifies that you have passed selection. Moreover, anyone wearing that badge has been to the bottom of their soul.

The selection phase of our training was not just a process to see the qualities of what a potential candidate has to offer the unit. The hardship and intensity of the training brings you to a place where you are completely vulnerable, both as an individual and as a group. This creates a baseline factory setting and makes you rely on your buddies, creating not only a bond and trust that is immense but also an unrivalled understanding of what a team really is.

◆

A short break to recuperate, eating what felt like ten meals a day and sleeping 20 hours a night, was followed by the beginning of the rest of our lives.

Our small group, known as Golf One, would now begin the process of learning how to become a Special Forces operator. Most of what we are taught is a trade secret so I am unable to go into a lot of specifics on the training, but the course was a huge step up from anything we had done before. I would spend weeks being trained and tested on all the qualities of an operator: marksmanship, decision-making and the other things in the tool bag. It is tougher in ways than selection because the pressure to perform and meet the standard is always there. The ability to perfectly and repetitively perform a simple task lies at the heart of Special Operations, and it's anything but easy.

I had a dual existence at this time: the operator and focused individual during the week and on weekends, if free, the heavy drinker. I have some really good, old friends, Foley and Keller, who took turns putting up with my shit – excused on account of my marriage break-up. I developed an alter ego during those years – the guy who was angry at the world after a few pints – but I managed to keep it separate from work.

———◆———

Over the next few months the training focused on individual and team skills and the ability to think and quickly make accountable decisions. We learnt to not be afraid of making mistakes, to admit failure and to correct and learn from it. I also learnt not to take things personally, especially others' mistakes and criticism.

Our course sergeant was a really calm and relaxed individual who would hardly raise his voice, and he had a great sense of humour. He was the type of leader I was drawn to and I worked on becoming like him. We would spend weeks on the range, constantly firing all the various weapons until we were marksmen on them all. The attention to detail on shooting is taken very seriously and we each had to keep a detailed, daily log of our progression, or lack of it. This is a good idea, as it highlights patterns and weak points to allow extra practice in these areas and can be used for any skill-learning process. The score is not counted up – rather it is counted down from the maximum and calculated from the points you 'drop', or lose. That was the attitude and mind-set instilled in us from the start: not scoring 75 out of 80, but dropping five points.

I learnt so much from my instructors. We would be part of a unit where a code of honour and behaviour is lived with absolute adherence and without question – and no one dares to break this code, ever.

It was also the first time I was exposed to what is known as the 'debrief sandwich', a simple yet effective method to critique the performance of an individual or team. Instead of being criticised, which can be discouraging, in a debrief sandwich the team is allowed a chance to express how they felt things went and what they would change. The debrief then begins by delivering a positive message of observations on what went well. This is followed by the constructive critical point, delivered in a measured manner to identify the important part to be corrected, and in no more than three points. The final point is again positive, on how best to correct the mistakes and this is reinforced in a positive manner. This will send the team away with a positive mind-set to specifically improve what they are attempting to do. It works well and stops instructors – or managers in civilian life – from spending too much time on a debrief.

———◆———

The final phase of our skills training is all geared towards anti-terrorist training and hostage-rescue operations in particular. This is another steep learning curve. Up to now we had been just mainly dealing with bad guys, but now we were dealing with complex situations with multiple, unarmed friendlies shielding the bad guys.

We would train to assault various buildings and rooms, as well as vehicles and aircraft in multiple scenarios. As a team we learnt how to force entry with mechanical or explosive means and neutralise the terrorists or 'tangos' in the room. The team would pause in a corridor and clear each room as they moved along. As the lead buddy pair you would be the first two men into a room and would be prepared as such. That moment when you are formed up outside a room with your team, as the first man you control the exact moment of entry, after a simple signal from your buddy, telling you he is ready. You wait there, taking a deep breath and composing yourself before you make that step in the door, because once you go, there is no coming back. You get in that door and deal with the main threat as soon as you see it because your buddy is on your shoulder. You have to get into the room as quickly as possible because the entry point is usually where you will take fire – it's known as a bullet magnet for a reason. When you decide to go you must commit fully and dominate the room.

The skills exercised in this type of training are aimed to increase reaction, which needs to be fast and immediate. Hesitation means death. We carried out training in our 'kill house', a building where we practised using both live fire and 9mm paint-conversion kits on our HK MP5 sub-machine guns and Sig Sauer 9mm pistols. The paint ammo allows a live enemy to shoot back, considerably increasing realism. We were subjected to constant training in multiple situations – whether to shoot or not, to kill or incapacitate – over and over, to a point of muscle memory action and instinctive reaction.

The confusion of a house assault is often extreme, with various scenarios, explosions and smoke to be negotiated. A respirator and night-vision goggles further hamper your reactions. You have to be present at all times. If you need to cover a door, that is exactly what you do without even glancing away from it because the second you look away it opens and you're dead. A quick identification of the risk is crucial.

I have developed a simple and effective process for situations when a quick, split-second decision needs to be made. Firstly, a number of factors affect the quick-decision-making process:

- ► **Confidence**. Your own, and how prepared you are to complete the task.
- ► **Ability**. Your skill and that of your team and the assets you have to do the job.

- ► **Time**. For quick decisions this is critical.
- ► **Opportunity**. The ability to act may be restricted to a window of time.
- ► **Elements**. You and your team's condition and whether you're tired, cold, wet, hungry, uncomfortable or distracted. This can also include what is going on around you at the time.

Once you have considered these factors, you move onto action, and in order to make a quick decision, you have to assess the following:

- ► **Situation**. Read what is happening.
- ► **Risk**. Immediate to you / your team / your goals / your plans.
- ► **Action**. What you need to do immediately; your course of action.
- ► **Commit**. You and your team must go with it fully and carry out the decision, right or wrong.

I have followed this process countless times in training and on operations – it works, as you will see on further reading, but remember the process is designed for quick decisions, often made by one individual. Learning this was invaluable to me, but I also found out there are times when you have to make decisions as a group. The last part of the ARW training would teach me that.

———◆———

We knew we were close to the finish but, in true ARW style, you are often kept in the dark or fed misinformation to keep you on your toes. Our training would be condensed to allow us to be part of a forthcoming deployment for the unit, so some training was cut until after our return. The day we were finally given our green berets came as a complete surprise and followed on from a tough battle run, carrying kit to a finish point on the Hill of Allen in the midlands. The Hill of Allen is where Finn McCool and *Na Fianna*, the elite mythical warriors and protectors of the high king of Ireland, were based in ancient times. The unit is linked to this band of warriors in tradition and spirit. In fact, some of the tests we still undertake can be traced back to a variant in ancient times.

That day, on our arrival at the hill, the whole unit was waiting for us. We were presented with our headdress by the unit commander, who welcomed us to the community in a simple ceremony – no secret handshake, we're not the Masons! It was such an honour to finally wear that green beret. Yet the moment was bittersweet because I knew that, despite my pride, and all the effort, training and sacrifices I had made to get here, it had probably cost me my marriage.

◆

We were now on call 24/7 and this meant that we had to be prepared to deploy anywhere in the country in one hour and anywhere in the world in 96 hours. Our main role was in an anti-terrorist capacity, which saw us as the top-tier unit for hostage rescue on land, sea and air. We were constantly in a state of alert, ready to go at short notice, ourselves and our kit and weapons packed and ready to go.

Despite my professional success and commitment to the SF, this was a difficult time for me. Just before Christmas 1999 I went back to Cork and met up with a few army buddies for a pint to celebrate the season and my getting into SF. I had a lot to drink and when the pub closed I jumped into a cab and without thinking gave my old address, where I had lived with my estranged wife. I arrived at the door of the cottage half cut and opened the door with my key to be greeted by herself and her new boyfriend. I was still paying the mortgage so that's when I snapped.

I threw a few digs at him, nothing of great effect as I was worse for wear and they both had the good sense to leave in a hurry. And me? Well, I went nuts. I began by smashing all the windows by jumping through them, and demolishing all the doors. Having wrecked the house, I moved on to anything I could find in it, until I noticed blue lights outside and someone with a torch and high-viz jacket at the door – or where the door used to be, I should say.

I was very lucky that the gardaí didn't come in, as I would have gotten myself in real trouble. Luckily Margaret had called my brother Dan, who took me back to his house. They didn't press charges, since it was my own stuff I was breaking. At least nobody was hurt besides myself and my wallet – I had to replace all the windows and doors. This was days before Christmas so I got a guy to repair the damage so Margaret could move back in. I spent that Christmas at my mam's house. I remember it was a pretty miserable one.

I was at the end of my tether. I had been completely wound up for months, pushing thoughts of my situation out of my head, hence my aggression after a few pints. I didn't acknowledge how I felt all along and I didn't deal with releasing the demons in a controlled manner, so they all came out at once and of their own accord. It was a wake-up call for me: you can't just suppress negative feelings and pretend everything's fine. I had, for the first time in my life, completely lost control of myself. I couldn't let that happen again.

———◆———

Once more, in the spring of 2000, the ARW would save me. I was going to be part of a task unit that would deploy to the jungles and bushland of East Timor in a special operations task unit (SOTU), training to deploy to the Indonesian archipelago to counter an insurgency war.

We trained and prepared for a number of weeks before deployment. I was trying to soak up everything I could from more senior guys in the team and impress the team leader (TL) and 2I/C.

East Timor had just been given independence from Indonesia, which still controlled the west of the island and sent insurgents there to cause trouble. We flew in from Australia on a C-130 Hercules transport aircraft – not known for its comfort – and landed on the shortest jungle airstrip I have ever seen. I've been on bigger treadmills. It was a sharp descent and hard landing to say the least and the ramp opened to a blast of humidity that knocked back a planeload of hungover Paddies after our last blow-out in Darwin, Australia, the night before.

To add to our misery we were greeted by a squad of Maori warriors of the New Zealand Army doing a Haka to welcome us. It took all of my composure to stand straight and maintain eye contact, but it was incredibly impressive to see these guys stripped to camouflage jungle trousers, performing this ancient ritual. It cemented a lasting bond between us and the Maori soldiers that was very special – later, we even taught them hurling!

We were relieving another task unit in the highlands and soon began to settle into our new environment, high up and often above the clouds. We patrolled along the rivers and jungle tracks to counter the militia, who would infiltrate the small villages along the border close by and terrorise or kill the villagers. It was tough going, fighting through the undergrowth and bush carrying the kit, up and down the ridges and valleys in the rough terrain.

We were supported by Bell Huey UH-1 choppers and the first time I got picked up by one and flown over the jungle was such a buzz – waiting in team formation, hearing that distinct *wap wap wap* sound of the rotor blades as they came up the valley to an extraction point. The downdraft from the blades as it flared for a quick pick up and lift off seconds after we sprinted to the open doorway; sitting along the side, legs hanging out looking down on the jungle canopy below, with the warm wind in your face and the sound of that machine. It was a big moment for me. I had read all about the Vietnam War where these iconic helicopters had been used on such a scale, and it felt surreal and exciting.

It became less surreal very soon as we slogged through the jungle, everything there either prickly or wanting to bite you. The heat and humidity were energy-sapping. I was the team machine gunner so carried a Minimi light machine gun as well as extra ammo in five or six 200-round boxes, so I was well weighed down.

It was monsoon season and the rain, when it came, was like standing under a powerful shower – everything got soaked and ended up rotting very quickly. The positive side to this was that when you needed a wash you just stood outside with your soap, which we sometimes did. My team was often on operation on various tasks and we would have support if it could fly in to help, but in a lot of situations we were on our own. This was great as far as I was concerned because it meant that our interdependence was extreme. We had an outstanding bond as we could stick to the simple basics of soldiering. My team leader was a hugely experienced and methodical man called Fred. He understood very quickly the strengths and weaknesses of everyone on the team, which he used well. He was a little bit nuts though, which is always a good thing, but he was a great team leader to me on my first SF deployment.

When preparing for an operation the TL and deputy TL would construct a large part of the plan but also include input from the four team guys. This collaborative approach to making a decision is the norm in the unit at all levels and gives the best options for a plan. The leadership won't see everything, so fresh eyes and input are always welcome.

I found this a good confidence tool for me as Fred would instruct us to carry out individual tasks, which would immediately have us buying into the communal plan. This means that all the team has ownership of the mission, not just the leader. Obviously, the TL has the final say in what happens and for snap decisions there is no time to canvas opinion. In a small team, when the shooting starts only one man should manage the fight; everybody else reacts to direction.

We applied a specific process to develop our more deliberate plans in relation to the situation we were about to enter and our ultimate aim in that situation. In military parlance, it is known as an 'estimate of the situation': This is a format to get you thinking about what is in front of you and why, importantly guiding you to the best effect you want to achieve:

▶ What is the opposition doing and why?
▶ What do we need to do?
▶ What effects do we need to have?
▶ Where can I best accomplish each effect?
▶ What resources do I need?
▶ What control measures do I need to impose?

This method worked for us because we could rely on our fellow team members after we sized up a situation and came up with a plan. Our bond remained strong even under the most difficult of circumstances.

———◆———

The bushland and jungle was home to a lot of nasty creatures: the taipan, for example, is one of the most poisonous snakes on the planet, with highly toxic venom and a very aggressive nature. They were, of course, also all over the gaff. We didn't carry anti-venom so if you were bitten in the arms or legs, you'd have two hours. If you were bitten on the torso or head, you'd die. (We didn't carry the anti-venom as it is only useful as a counter to a particular type of snake venom or bite, so the wrong anti-venom could cause even more problems.) We understood the risk and had a heli medevac on standby. Luckily none of our team was bitten during the mission.

The jungle was so thick in parts that we had to rely on the rivers and stream systems that crisscrossed the rough and mountainous landscape to navigate our way. We would patrol along or through these waterways, which would rise quickly in the rain and become a raging torrent of brown water that could easily wash away a team. If you had to wade through a deep pool at any stage you were guaranteed to pick up a few passengers such as leeches, who would start the size of a matchstick, drink their fill of your blood and then fall off, now the size of a fat sausage.

One of my toughest physical experiences was an infiltration or walk into a target area to conduct a covert operation. We had received intelligence that the bad guys planned to get into a remote village near the border to kill the village elders. We had an operation of various teams that would be covertly inserted near this village to set up blocking positions and ambushes, to prevent these guys from causing harm.

We were inserted just after dark with a lot of kit as we had to be self-sufficient for up to seven days. The distance to our target area was less than 4 km from our drop-off point but the ground was extremely difficult, with thick bush and steep ridges. The six of us patrolled into this area, beginning what would be the march from hell. We scrambled up and down ridges in the dark for hours on all fours – each step was a battle to push through the thick vegetation, which clawed and tore into us. Thorns stuck me in the face and body and a large spike of something even penetrated my jungle boot at the ankle – that *really* hurt. Every few metres we would come to a step in the ridge, which would involve a team effort to crawl up and get each man over it, reset and do the same thing again and again. I was worn out, just keeping myself going and wondering when it would end. 'Our House' was on repeat in my head. And this was just the commute to work!

It was so humid that we could hardly breathe and were under time pressure to be in position before first light the next day. In normal situations we would not undertake this trek in the dark, but mission-essential requirements had dictated otherwise. We finally got to our lie-up position (LUP) just short of the target area in the nick of time to set our perimeter before dawn. All of us were completely exhausted and we just spent the day resting up, two men watching, before we would get on the target the next night. The team 2I/C was probably the toughest man I have ever worked with anywhere, and his tolerance for all types of hardship was legendary within the unit. Nicknamed 'Taz' after the infamously tough marsupial Tasmanian devil, he was designed to survive in all kinds of adverse conditions and had a simple get-it-done mind-set to go with it. He later told me that patrol had been the hardest thing he had ever completed in his life, which I'm taking as a compliment – he doesn't give them often.

I learnt a lot about myself that night and my ability to push through. I now truly understood how the will to succeed, discipline, attention to detail and personal admin were so critical. I knew now why on selection you need to test candidates to their core; you absolutely have to have people that can be relied on, you can't quit on a real operation. Some of the other teams were either unable to infiltrate or were compromised during the operation, but we managed to see it through completely.

◆

We would patrol regularly to a number of villages that were close to our small camp. We had a bit of a rapport with the locals – they liked us there as we prevented a number of attacks and had bagged a few bad guys. We also held medical clinics for them and helped however we could. They were a hardy people and I remember one of our medics stitching up a little girl, no more than four or five, with a serious gash on her leg. Her father held her in his arms as she underwent the painful procedure without even a flinch. These people had precious little in belongings and led a hand-to-mouth existence, living in mud huts with thatched roofs, yet they had such a positive outlook.

I always carried a couple of lollipops in my kit, not because I'm Kojak but as a morale booster – if things were difficult or I needed a moment to think, I'd break one out. It would settle me, give me some time to reflect. I ended up giving the children in the villages any sweets I had on me, of course. I also took home some valuable lessons from them, because how they live and how uncomplicated their attitude was would give anyone perspective on their own lives. The children also had their own edible treat of sorts, a large flying beetle, which they would grab and wolf down immediately like a child here would eat a chocolate bar. We called them 'Kinder Beetles' – like a Kinder Egg but without the magic inside.

We bought supplies for the villages and in particular for the tiny schools. Blackboards, chalk, paper and of course footballs and the like. It was so humbling, it just lifts you up. The people were so grateful, especially with the volleyball sets we got them. Every time one of our patrols passed through we would play a few games, usually against the whole village at once.

We supported the Kiwis in an operation later where a village had been attacked and we were trying to flush out the militia from the jungle, who had attacked in some strength, so were up for a fight. They didn't find us but hit a smaller Kiwi team and engaged in a close-quarter contact in very thick jungle, where the Kiwi team lost a man. They found him soon after when they fought their way back. He was dead, finished off by the militia as he lay wounded on the ground. A stark and powerful reminder to us all that snakes weren't the most lethal creatures in this jungle.

Despite the horrors and the hardships we experienced in the jungle, the Timor experience was probably one of the highlights of my career. It was all about the basic skills of Special Operations soldiering and the bond of a small team, living and working in the harshest of environments. There was no cake, it was all bread-and-butter work, but if you are with the right group of like-minded, performance-driven individuals, nothing is ever a problem.

CHAPTER 5

AWKWARD

The summer of 2000 saw me back in Ireland, in a training environment in the ARW, where we had to catch up with some of the more specific areas of training that were put on hold for our jungle mission. One of these was parachuting.

We completed the preliminary training and ground-school phases of the course and were put in small groups of four, called 'sticks', ready for our first lift and jump. I'll never forget that first time. We jumped from a ridiculously small Cessna aircraft. The space was extremely tight inside the aircraft and as the first man out, I was seated in the open doorway. With the pilot at my right shoulder I was sitting in what is known as the 'dentist's chair', which is a purpose-built seat, facing backwards to the direction of travel.

The tiny aircraft revved up and shot down the runway in an attempt to gain enough speed to lift off with the payload, the wingtips dipping from side to side. I am not a nervous flyer but going in that thing brought it out of me and I could feel the anxiety rise as the Cessna climbed in a circle to jump height.

I didn't feel good at all. Was I scared? I started my coping techniques by taking deep controlled breaths, using a buzzword and rationalising what I was afraid of, practising the movements as I sat in the doorway. As the instructor gave the signal to get ready, my anxiety stopped instantly. I stepped out the door and edged to the jump spot, which basically meant hanging off the wing, with a foot on the strut. He gave me the 'GO!' and I let go, dropped with the sensation of a rollercoaster and carried out my drills to the sight of a lovely full canopy opening above my head. I identified my target below and landed well, standing up. Happy days.

I told the ARW jump master afterwards about how I felt before I jumped and that the fear had gone as soon as I started action. He laughed and told me that it was the dentist's chair that does it, the position you are in travelling backwards. The perspective often gives people the horrors. I thanked him for putting me in that seat, then proceeded to do five more jumps that day without much anxiety. Parachuting is a large part of unit operations, so I had to get used to it quickly, which I did. I'm no sky god, unlike some of my air-insertion specialist friends who will jump from a huge height with oxygen masks and land on a target the size of a snooker table. I did what I had to, but I was never a fan.

Fear is a powerful emotion that gets your body ready to deal with danger. We all fear different things: failure or being judged, for example, which manifest themselves as worry, anxiety, phobia or terror. The 'fight or flight' reaction dictates whether we go forwards or backwards, as fear can stop us from attempting a lot of things and affects how we think and make decisions. All of this can be controlled if you know the source of the fear and the mechanism that sets it off in you: in my case it was sitting backwards in that tiny aircraft. Yet fear can be managed, controlled or reduced to a point where you can continue with your particular task. Fear is not a negative if it is controlled and keeps us focused and sharp.

These are my ways of coping with fear:

Face the Fear. Get used to what scares you. Practise parts of the task to take the sting out of it, building up a tolerance in the process. The more you put your hand in the fire, the longer you can keep it in there as the mind becomes used to it.

Breathing. Take slow, deep controlled breaths, in through the nose for two seconds and out through the mouth for four seconds. This will oxygenate your respiratory system, slow your heart rate and calm you down.

Buzz Word. Think of a word and repeat it in your head or say it out loud to keep the brain focused. You can also use the happy-place concept by simply visualising being in a place that comforts you.

Take Action. Having an active function or the focus of doing something positive during the experience will reduce fear, giving you something to act on.

And always remember that courage is not the *absence* of fear – it is the management of it by using coping techniques and completing the task *despite* fear.

———◆———

For the next selection course I was appointed DS, mainly because I was a bit older and more mature than the other guys who had come into the unit with me. (Old and young are often paired up in instruction, so that younger or less experienced DS are mentored by experienced guys.)

The course began the first night with the usual mayhem, although it was a new experience to be pushing out the madness as opposed to receiving it. Some guys quickly forget what they had gone through as students when they are promoted to DS and can sometimes be unrealistic in what is expected. I was still so new that I almost reacted as a student and joined in with the push-ups when another DS shouted the order!

Just because you get through the selection process and serve in the unit doesn't make you infallible and all-knowing. Unfortunately – and thankfully rarely – the occasional asshole slips through, only becoming a better-qualified asshole after the process. I've drawn up a list from my own observations over many years of the main categories of recruits:

> **The Fuse**. Anxious, all over the place and running around like a ticking bomb ready to explode when the moment arrives. These guys don't last too long.

> **The Grey Man**. Quiet and composed individuals who work enough so they don't draw too much attention to themselves.

> **The Hero**. The one who will jump into everything, help everyone and take control of whatever they can at all times. These types are absolute believers in the organisation and themselves.

> **The DS Watcher**. Will perform when they think that the DS can see them, but when alone with other students they are not as forthcoming. They are basically putting on a show for the instructors.

The Team Player. The guy who will work hard at all times for the benefit of all.

The Walter Mitty. Guys who want all the fancy gear, guns and the fantasy without having thought about the hard work. These guys look the part but are out of their depth.

The Natural Leader. Unites all the group by providing example, support and direction.

The Mé Féiner. Look out for themselves and don't help or bond with others.

The Bluffer. Does exactly what it says on the tin.

The Slogger. The no-frills guy who puts his head down and gets through.

Selection puts people in an extremely intense environment, where a lot happens in a very short time. It quickly strips away the protective layers so they can no longer hide behind them. The main battle on a selection course is the one the students have with the demons inside their heads, while our job as a DS is to ensure that battle commences. But it's not just in the army – these people exist in every office, company, organisation and business I have been involved with.

As a DS, it is critically important to understand how empathy works and how to act on it. My own emotional intelligence is good, and I am able to relate to and create a link with different people very quickly.

The selection course tests various qualities but if you just keep putting the boot in students then nobody will pass, and that is not the objective. Being able to use cognitive empathy, to step in their shoes, will keep you in touch with students as individuals. This will allow you to tweak events to determine whether you want to up the pressure or turn it down a little.

I was testing a student who had just started a covert recce (reconnaissance) of a target area in the mountains but was making a balls of it. He was a strong young officer – a classic hero type – who had ticked all the boxes thus far, but here he was under pressure and having a bad day.

I could have just let him carry on with his recce and watch the unfolding disaster. Instead, I told him to stop what he was doing, to take off his helmet and backpack and sit against a tree. He did so and looked at me, bemused. I pulled out a morale lollipop from my kit and gave it to him. He was absolutely in shock. I went on to tell him that he had as long as it took for him to finish the lollipop to think about what he was doing and how he might change things around. He sat there for ten minutes and had a think about his life, went back at the task and did a better job on the restart. He passed selection eventually and we ended up running a task unit together some years later.

We all need a little empathy sometimes, regardless of how brilliant we think we are – I know, because I got it plenty of times from good people myself.

---◆---

I was leaning towards the maritime side of the unit and wanted to complete a combat diver course, which was the entry-level course and a bit like a water version of selection. This element of the unit specialises in maritime counter-terrorism, ship or platform take-downs, boat handling, beach recon and tactical covert sub-surface dive-insertion operations.

The initial three-week course is run by unit dive instructors in Leinster, then moves on to four weeks in the Naval Service Diving Section (NSDS) in Haulbowline Naval Base, where my grandfather was born. The first part involved us being worked very hard on the basics of military diving. The bulk of it is completed in a mountain lake, so the water is pitch-black and cold. The idea is that if you get through this phase, the naval part is more achievable.

The first phase took place in the weeks leading up to Christmas 2001. In this extreme cold, the punishment for mistakes is a skinny dip in the icy lake or canal waters, so you learn fast. I remember we were being inspected for attention to detail on some of our dive equipment by our sergeant and all students were pulled for a problem except for me. The punishment was a skinny-dip duck dive to the bottom of the canal to recover a weighted box, requiring a team effort. I was told to take a step to the rear but I togged off and jumped in with the team anyway. Sometimes you have to take a hit with, as well as for, the team.

Having completed the first part of the course, we arrived in Haulbowline on a Sunday evening in January to begin in the naval dive school. It was a freezing cold night with sleet falling. Our vehicle pulled a large trailer packed to the neck with all manner of dive kit as well as boats. The course was due to start the following morning, so we arrived early to stow our kit and get a good night's sleep prior to the start. The dive section is located right on the quay wall, so we pulled up in our army vehicles and there, standing at the quay, was a hardy bullet-nosed naval NCO, waiting menacingly for us in the gloom.

He was the dive chief petty officer, addressed as 'Chief', as tough as a barnacle and the man who would put us through this incredibly difficult course for the next four weeks. Before we even got out of the vehicles, he shouted out the immortal command 'Awkward!' – a naval order for a diver quick dress (to dress into dive rig in two minutes or less). We exploded out of the cars, climbed over all manner of equipment in the trailers and tried to grab our kit and dress as required. It was mayhem. Before we could get our bearings, we were in the tide doing circuits of jumping in from height into the dark harbour water, surface-fin swimming and climbing out, again and again. 'Welcome to the Navy!' Chief bellowed each time we passed him as we ran by and finned around the harbour.

The course focuses on the ability to pay absolute attention to detail because a lot can go wrong underwater, especially at night. You have to be absolutely surgical with the preparation of equipment and check every single detail prior to going in. The physics, medical and dive regulation lessons on the course are detailed and intense and 95 per cent of the course is practical, with the book work done in your own time. I should probably clarify that combat diving is different to what I'm sure you have in mind – it's anything but warm blue water and cute little Nemo-like fish all around. Combat diving is a method of insertion into often hostile territory. It's conducted mainly at night, usually in the freezing cold and in water so murky you can't see your hand in front of your face.

To increase your endurance and get the most use of your air tank, or set, the Navy uses what are called jackstays: large ropes, one sunk and anchored to the bottom of the naval harbour and one secured in the tidal channel. The jackstays are temperamental – the instructors put us in when the tide is at its strongest, so we end up hanging off the rope at 90 degrees, inching hand over hand as we are pushed by a relentless tide.

In buddy pairs we would sink to the bottom and swim along this rope, like a handrail stretching for hundreds of metres, for hours on end. We could only come up when we reached the safety limit of 'low on air', and even then they'd change the set on your back while you hung off the small boat so you didn't even get out of the water. The chief also had a twin air set on the boat, which is double the twelve litres of compressed air normally used, as a threat to anyone who wasn't getting the maximum of 70 or 80 minutes per dive. It was called the Ice Maiden, and donning it would mean being under for two or three hours in one evolution. After some dives your hands were so cold that you'd need help taking your kit off.

The bond of a team and in particular a buddy pair is critical: you have to watch your buddy and check him as much as possible because your lives are literally in one another's hands. With no underwater communications, all chat was with hand signals (or by touch at night), which creates a sixth sense between dive buddies. The naval dive instructors worked us very hard but there was an unspoken respect between them and us. They are ultimate professionals and outstanding at what they do, passing those skills on to us with openness and honesty – they make us earn it, though.

The days were long: diving from light to darkness, then night dives, followed by hours of admin, filling air sets, sorting kit, mending holes in dive suits with bicycle repair kits and rigging dive boats. We had to prep everything needed for diving operations, so we had to allocate jobs and take turns doing the worst tasks. Collaboration was key as we were constantly wet and cold, with our hands in pieces from the salt water and rope work. That understanding of the work ethic required by all is a constant in the maritime unit, regardless of rank or position. It's a relentless taskmaster but it prepares us well for working in the sea, which is unforgiving and doesn't give second chances.

◆

The toll on the body is high on the combat dive course. You become run down from close to two months of constant hardship of being constantly wet and cold, spending hours in the water. I picked up a stomach bug, which saw me vomiting for 36 hours, even underwater. I had the timing perfect as to when to take out and put back in the demand valve in my mouth between nausea bouts so I could breathe. I also got haemorrhoids which, yes I'm going to say it, were a pain in the ass, mainly because they bled profusely underwater. (I got these conditions mainly because my immune system was down from months of overexertion and lack of recovery.) My big problem was trying to keep them clean so they were not infected, since reporting sick would mean I would be ordered to leave the dive course. To counter this, I would spray them with an alcohol-based antiseptic aerosol – directly onto the target area, every chance I got – which stung like hell and would bring a tear to a glass eye, but kept them clean. It also entertained my dive buddies no end with the sound effects.

We learnt how to compass-navigate underwater and use specific tools to allow us to get from A to B. If you missed a target or misjudged the tide and got pushed away, the chief had a number of ways to highlight your crime. A favourite was to get the buddy pair who messed up to walk across the bridge onto the naval island and in the main gate of the base, fully dressed in all their dive kit as the sailors or 'sea hags' laughed and slagged us as we went by with our camouflage dive suits and masks on. Like every other type of challenge, you just have to break it into small bits and you will get through it, gradually increasing your tolerance to stress and resilience as you go.

———◆———

Having finished the naval training and after spending a couple more months on different exercises, I was now coming up to three years in the ARW. It was an intense, hard-learnt and earned way of life, I was finding out. It's not just a case of get in there and then it's Oakley sunglasses and fancy kit after that. I loved that you couldn't hide behind a rank or position and were challenged constantly by people who call it like it is, giving you the truth with both barrels. It was difficult for some army officers in particular, who came from a place where rank structure is very rigid. The ARW was more flexible and an officer would be called to task by his teams if his plan or input wasn't good enough. The division of rank in SF is fluid and means that the person who is higher in rank doesn't always have the final word. At several times even at the senior rank you fall under the control or command of lesser ranks, who lead because they are more experienced or specialised. A buddy pair could be made up of a captain and a corporal who share everything together with no barriers or pretence of position. Some struggled with that and after being highlighted by NCOs would resent the fact and not have a good tenure in the unit. The good ones embraced this system and excelled in it, understanding that it produced a higher performance and more honest working relationship.

I had completed a long-range reconnaissance (LRRP) course at this time and again it was a very tough and rewarding experience, with more patrolling and carrying of heavy kit, but this time even further. The course is a direct window into the practicalities of preparation for gathering information on a strategic level and turning it into intelligence and the process that entails. It shows how important the correct information is at the appropriate time and how information constantly changes, which is true in any business.

———◆———

It had also been a tough time in relation to my break-up with Margaret. Despite having been married for fewer than two years, our break-up was protracted and messy. A circuit court appearance to legally separate was a disaster as I had not prepared for it. It left me in shock again and kicking myself that I assumed it would be all fine. I appealed to the high court and this time prepared my case in detail with my legal team, as I would in a military operation.

This time we managed to reach an agreement outside of court. I was relieved and happy to move on and be done with it. I needed closure. I'm not blameless and I'm sure it was hard for her when I was away, but she never said anything. Once again, I learnt that military skills apply to civilian life, and when you prepare correctly you give yourself every chance to succeed. I could now draw a line under this and move on with my life.

———◆———

I spent the winter and Christmas of 2002 in Kosovo in the Balkans, where temperatures dropped to below −20 in the mountainous landscape. The country was just beginning to recover after a brutal occupation by Serbian forces.

I was deployed with the NATO-led mission there, attached to a regular army unit for the six-month mission, and I was glad for the change of pace from the ARW. After three years of intensity I needed a break. Not long before I deployed, I met an amazing girl, Sinéad. She was full of life, and I realised that I could finally be happy with someone. We hit it off from the start. She was the piece missing in my life.

The outfit was a transport company that provided convoys for NATO on missions through the mountains of Macedonia to Greek ports. The ancient, rugged landscape had hardly changed from the time of Alexander the Great. My task as part of the security team was to provide protection, along with a group of 20 others. In addition, as a fully qualified PTI, the CO also employed me to manage a well-stocked gym in our camp.

Here I ran various training classes as well as individually training most of the 120 members of the company. They came in all sizes and shapes as well as attitude. The whole experience brought me back to dealing with different people again – not just like-minded SF types – which is a great thing because it keeps your feet firmly on the ground.

I also carried out a number of close protection (CP) or bodyguard duties for visiting Irish dignitaries from the military and Department of Foreign Affairs. For these I hand-picked and trained up a number of the security team to conduct these operations.

One such visit was from a senior Irish officer, RH, who was coming to be an expert witness in a Hague Tribunal war-crimes case. He also just happened to be the first CO of the ARW in 1980, when it was founded, and had a large part to play in the transition to the unit that exists today.

With my team I picked RH up from the military airport in Pristina. I had heard the stories about this man: his name was almost folklore in the ARW and he was revered by all the older guys who had served with him. I stood at the exit door and as I saw him approach I greeted him with a salute. He immediately shook my hand. I could see that he was delighted to be met by a Ranger wearing the green beret, which he had inaugurated. It was a moment of great pride for both of us.

In two marked Irish KFOR (Kosovo Force) vehicles we travelled to Mitrovica to the north of Kosovo, where RH would be the expert witness against a Kosovo Serbian colonel being held there. This would be a show trial for NATO in the town, with its iconic bridge across the River Ibar splitting the warring populations of Kosovo Albanians and Serbs. The area was a flashpoint for both populations during and after the war and was now, with the trial in the French-controlled area, at the centre of worldwide attention, with daily protests and rioting. The trial was being held in the former magistrate's court in the Serbian side of town. Large numbers of KFOR troops were stationed all around the area since the threat of a terror attack on the location was very real.

I sat at the back of the court and listened to some of the incidents this Serbian colonel was being accused of. It was intensely hard-going: burning whole families alive in their basements, summary torture and executions of Kosovo Albanian men. He was also accused of setting up what were called 'rape houses', where Kosovo Albanian women and girls were held as slaves for the pleasure of his troops, and razing to the ground religious sites, houses, hospitals and civic locations in an attempt to completely wipe out the culture of these Muslim people, with a hatred that went back to the days of the Ottoman Empire. The Serbs famously halted the advance of the invading Ottoman Turk Army back in the middle ages on a site now famous for all Serbs, 'the field of crows' in Kosovo. The reclamation of this area was amongst the reasons given for invading Kosovo in 1998 – the contempt these two groups had for one another went that deep.

During a recess, RH was invited to a small impromptu meeting with the main defending Serbian counsel in his chambers upstairs above the court. When you are a bodyguard, you don't like a change of plan, or 'fast ball', since it often means you have to go somewhere you haven't checked out beforehand. I wasn't happy but RH insisted, so we climbed the stairs to a small office at the rear of the building.

RH was a tough man, well able to handle whatever was going to happen, but I just didn't like the extra risk. I had a very quick look into the tiny office, in which sat two Serbian lawyers, so I agreed to sit outside the room if the Serbian bodyguard did the same. We sat directly across from each other outside the door. He just stared at me with contempt, so I stared right back with a smile.

The meeting ended and we walked back to the main courtroom where RH took the stand and delivered a measured and professional testimony of his experience in that phase of the conflict. The Serb colonel was found guilty of some but not all of the charges and later served a prison term. I brought RH back to the airport the next day and he travelled back to Brussels where he was based. He retired soon after and I have met him on a number of occasions at ARW reunions, where he often remembers our day in court in Kosovo. Sometimes you should meet your heroes.

———◆———

For New Year 2003 I travelled home to surprise my mam, who hadn't been well. She'd had a stroke the previous year, but thankfully fully recovered. I spent most of the leave in Cork, with a good part also in Dublin with Sinéad, whom I brought home to meet my family. Mam has always been a huge support to me, particularly during all the trauma of my marriage breakdown. She was direct and honest, and always gave her full support in whatever way required. No matter where I was in the world, I made time to contact her. Nobody has your back like your mam.

I returned to Kosovo, finished the deployment and brought my mam to Paris for a long weekend that spring. It had been her dream for as long as I can remember. I pulled out all the stops, taking her everywhere she wanted to go in the best style I could. It was nice to do something for her for once and we both had a brilliant time. We sat one morning in Notre Dame Cathedral as a choir of children sang in the nave. It was so moving that my mam wept with joy listening to the angelic voices, holding my hand without saying a word, a moment I will never forget. This experience in Paris reminded me that you should always, when you can, do the things that are important with the people who are important to you.

———◆———

My next course was known officially as a Dispatch Riders' course (DR), which would have me learning to ride various types of motorbikes, both on and off road. Back in the day the army used bikes to deliver important dispatches, although now we use them for discreet surveillance operations, direct action and for reconnaissance in a conventional warfare role. The course was more of a stuntman's training, because that was what it involved – white-knuckle, death-defying stunt riding.

I had never been on a motorbike – bar riding on the back of a Honda 50 – so I wasn't sure if I was suited to it. The first week started at a walking pace, learning motorcycle control. Then, with a small bit of knowledge – always a dangerous thing – came the road phase, which saw an ever-increasing amount of speed and danger as we drove around the country getting to grips with throwing the bikes around. My confidence was on the rise as we moved into the off-road phase, when we transitioned from road to dirt bikes. This part I really enjoyed and loved the thrill of cross-country and learning how to ford rivers and basically jump over everything. The attrition rate of the course was high enough, as expected, with several lads getting injured with various breaks and bangs along the way and my good friend Baz busting his shoulder on the off-road phase.

I had a close call or two myself, but I really only understood how lucky I was when we were conducting a follow-the-leader exercise one day. I was following Danny (the same DS who had spoken to me years before when I had failed my first selection course). He led us along open ground and over mounds and hills and we all had to keep up the pace and stay with him.

At one point we stopped and watched as Danny demonstrated how to go up and over a large mound, and control a descent down the other side. I was next so decided I would give the bike a good rev to get up the approach side of the hill, but instead of a controlled descent on the other side, I just flew off the top, Evel Knievel-style, like it was a ramp.

As I launched skywards I looked down at Danny, who I was sailing over at the time, contemplating how I would be killed on the impact that was coming. The ground came up very quickly and by some miracle I managed to land the bike on the back wheel and continued on to where Danny was, pulling up next to him. He looked at me and said calmly, 'Ray, don't do that again.' I'd had a lucky escape and learnt an important lesson: stay within your limits until you build up enough experience to ensure your survival.

———◆———

On the home front things were going very well for Sinéad and I, so much so that we decided to move in together and buy a house. This was a brave move on my part because I wanted to leave my demons behind, which I did as the angry man was now buried, and move forward with my personal life. It was also taking a chance for Sinéad, who accepted completely who I was. She had no airs and graces about her, or the life she wanted. I had to de-Dublinise her a bit now that she was living out in the sticks, but she is a brave soul and had as much if not more courage than I did.

CHAPTER 6

HOTEL AFRICA

n November 2003, a week after Sinéad and I moved into our house, I was deployed to Liberia in West Africa. Liberia had descended into chaos, having been fought over by various factions, mainly along tribal lines. Peace was being enforced now by a UN-backed government left behind by the despot leader Charles Taylor when he fled the country. The coastal and capital city of Monrovia was relatively secure, but outside the city was a different story and this is where we would operate. We had limited medical cover in the fledgling mission and the only medevac support we had was a Dutch medical vessel offshore.

We arrived on air transport and set up home in Monrovia in several small beach houses, a complex formerly owned by the disgraced Mr Taylor. We were the advance team – an Irish armoured infantry battalion would follow a few weeks later – and here on our own.

We were there only a couple of days when my team was tasked with carrying out a helicopter recce up north near a town called Gbanga.

As we waited on the helicopter landing site (HLS), three uniformed Americans arrived and started to chat to our CO on the far side of the HLS. They would be travelling with us, so what we thought was to be a heli recce would be something else entirely.

We were picked up by an MI-8 helicopter, flown by Ukrainian pilots in sandals and smoking cigarettes, and were briefed on the real operation – six of us protecting three yanks negotiating the surrender of some 2,000 rebels from the Model rebel group – landing an hour later on a soccer pitch on the edge of town. We spotted the pitch immediately, as it had several hundred people around it. *Is there a match already on here?* I thought to myself. Of course it wasn't a match. I could see the weapons, rocket-launchers and gang-style headbands of the rebels surrounding our HLS.

We touched down and we stepped off to secure the HLS in a circular pattern in the face of the young rebels who ranged from around nine years old to early twenties. They looked like a big street gang, standing there in silence as we took defensive positions.

It was tense and we expected it to kick off at any moment. A lot of these kids were on drugs and unpredictable as a result. I wasn't mad about the idea of having to shoot any of them, but my training would dictate my actions, as a threat is a threat, although I think they were more interested in watching us. Despite their bravado, like young kids around the world, they were also intensely curious and were happy to just watch us. The negotiation team proceeded with its task, which was all completed very quickly, thankfully. We re-boarded the helicopter and flew back to base without incident, the moral of the story being to always expect the unexpected. It's also handy to have all your information to hand before you set off, and not once it's too late to do anything about it.

◆

We shared our beach house with another team – and when I say beach house, I mean it had been one once: all that remained of it now were the walls and roof. Everything else had been stripped out. We slept on the floor in mesh mosquito nets in the incredible heat as we were still in the rainy season. The only comfort was the breeze that came off the Atlantic Ocean at night, but still it was near impossible to sleep in the oppressive humidity. Everybody suffered from various skin ailments due to the constant sweating from wearing our kit and body armour in the heat. Severe heat rash was common: sweat would block the pores, causing blistering and in some cases large red lumps, which felt like your skin was on fire. Of course the worst-affected area was the groin, with an extra little kick for some of the lads from something called 'swamp-ball'. I'll let you think about that one for a minute. Lucky for us we had a great doctor attached to the teams, Kevin, and he managed us very well.

It was also impossible to hydrate properly. The first weeks were like living in a sauna – as soon as you drank the water your body would sweat it out. The local water we obviously couldn't use so we had to rely on a fledgling UN logistics chain. This meant that water was rationed to so many litres of bottled water per day, so when yours was gone it was gone. I remember dreaming about glasses of cider with ice and driving myself mad at the thought of it, but we became acclimatised. My father had arrived in this climate 40 years before me, dressed in wool uniform … sweet Jesus.

The sea beside us was out of bounds because of the threat of sharks and pollution from the city around us. In addition, it wasn't unusual for bodies to wash up around our camp, so paddling was not recommended.

To add to the mix of heat, conflict and brutal conditions was the fact that we were all on Lariam, the infamous anti-malarial medication. Although it offered good protection from malaria and dengue fever (which could be fatal, and had been to some UN troops there), it had dreadful side effects, which ranged from headaches, nausea and emotional instability to psychotic episodes. A full task unit of us, armed to the teeth and trained to the highest level, given meds that cause episodes of rage and aggression – what could possibly go wrong?

Army policy at the time was Lariam for African missions, even though other alternatives were available. The Department of Defence in particular is still seeing the fallout of that medication, and hundreds of soldiers continue to be affected by it. It definitely impaired my own short-term memory, which never fully came back. I also had some incredibly vivid and crazy dreams, and not just about cider. I remember after I got home from the tour Sinéad had to wake me up one night as I leopard-crawled down the stairs in my sleep, looking for my weapon. When she woke me up with a prod from the sweeping brush, I didn't know where I was.

The TL of the other team in our hut was a guy called Derec. I got to know him the previous year when we both were students on a LRRP course. Like me, he had served in the army for a few years before entering Special Forces, so he understood how guys with prior service needed to be managed carefully. In the regular army prior to passing selection, I had climbed to a position of junior rank after nine years. But when you go to SF, regardless of experience and rank, you started at the bottom once again, which has its challenges. We discussed this at length.

Derec had a superbly calm way about him. He once told me that as a boy he stammered very badly – so much, in fact, that he slowed down everything in his mind to give the rest of him a chance to catch up. You would never think it, the way he conducted himself with such composure. He was also a good laugh and great to slag guys with, as we all did. I guess it comes with the lifestyle, humour as a pressure-release mechanism.

Derec and some other senior platoon guys went out on a recce one morning, a week into the trip, when their vehicle flipped over. Both he and our task unit sergeant, S, were seriously injured. The third guy in the vehicle was my TL, Taz, who escaped unscathed. He had been the team 2I/C in Timor and his robust nature probably saved him from serious injury.

S and Derec were worked on by our own medical team on site, then medevaced to the *Rotterdam*, the Dutch naval vessel with its onboard medical centre. The CO got us all together that night and announced that Derec had died on board from his injuries. It was a hammer blow to us all. S was still fighting to survive, which he did thanks again to our doctor, Kevin, who had him flown home for specialist care. He lost a leg and had some other major injuries, but he is a tough nut and made it, and is still in the unit today, a testament to his resilience and courage, having been close to death several times during those weeks.

———◆———

We were devastated, both on an individual basis and as a group. These two men were major players in our task unit. The grief emanating from us was tangible, it was so profound. We were given a chance to call home by satellite phone. I could barely speak to

Sinéad and she could hear in my voice that I was deeply shook.

We gave Derec the send-off he deserved before he was flown home to his family for burial. The rest of us just had to get on with the task in hand and carry on with the mission, which we did immediately. It's very much part of the mind-set that you get on with it, as the execution of the mission is the all-important priority, regardless of casualties and problems. The two men were a huge loss to us but after a quick reshuffle we got on with it, and that's the way it has to be.

---◆---

Purpose and activity are the best ways to deal with grief. It doesn't mean that you forget the dead, or that they aren't missed, but the time to deal with that comes after the operation is over. The army takes these type of events very seriously and a team was sent from home to talk to us and provide counselling and help. These guys are serving soldiers, so know the situation, are trained in the personnel support services (PSS) as councillors and are a great group of people, affectionately known to us as 'the care bears'.

Shortly after Derec's death our unit was informed that the UN would be commencing a weapons amnesty, giving a bounty to whoever came to Monrovia and handed over weapons, which they did. These we blew up. The exchange of dollars for weapons would see rebels line the streets and the project was going quite well until the money ran out. The rebels then took some UN and NGO staff hostage in the camp they used to hand over the weapons, so we had to go and sort out the situation that night. We managed to

Where it all began: Butlin's, 1977. I was standing on a milk crate to see the target.

In the middle, aged 16, with friends, Tom (left) and Trevor, in our FCA uniforms in 1988.

My passing-out day with the four people who set me on the path. Left to right: Jim, me, Mam, Mike and Dad. My nephews, John and David, are in front.

My passing-out march-past on 17 May 1990. I'm fifth from the rear of the nearest column.

In Parris Island with Foley after his graduation in March 1991, in a borrowed kilt.

Getting my UNIFIL (United Nations Interim Force In Lebanon) service medal in Tibnin, Lebanon, with friends in our 'whites'. Left to right: Gordon, Martin (RIP), Alan and me.

A short halt during a patrol in East Timor in spring 2000.

Sitting in the door of a 'Huey' in East Timor in 2000. I'm on the right – you can see my smile from a long way.

The jump position. Hanging from the wing of a tiny Cessna waiting for the 'Go' during parachute training in late summer 2000.

The best seat in the house: the thunderbox. On the beach in Liberia in late 2003.

Advance team. Mick and myself outside the German embassy in Monrovia, Liberia, in early 2004.

Women and children dance with joy as we arrive in Yekepe, Liberia, in December 2003.

Explosive-entry training during an anti-terrorist block in 2007. (*Courtesy of the Army Press Office*)

Fast-roping with my team from an Agusta Westland AW139 onto a naval vessel off the south coast of Ireland in 2008. I'm the one nearest the helicopter.

The 2009 selection course that nobody passed! I'm second from right in the front row, with John Killeen (Killer) seated beside me and Baz to the far left.

Clearing a beach, with Staff covering me, before conducting live-fire training on Bere Island in 2011.

Hostage-rescue training in 2012. (*Courtesy of the Army Press Office*)

My Malian platoon after graduation in Koulikoro in May 2013. I'm in the middle of my UK colleagues.

My viewing spot in the evenings in Koulikoro in 2013. I was longing to get into the Niger River below me.

Our mixed international team on attachment to the US Special Operation Forces, during the combined special forces event in Tampa, Florida, in 2014.

Hostage-rescue training as the platoon sergeant with a team at Irish Rail. I'm second from the right in the back row, beside the smallest ranger ever!

Christmas eve on full alert for an attack in Roshan Village, Kabul, with my Afghan girl, the beautiful Georgia.

Damage to the accommodation block and perimeter after the truck bomb in Roshan Village in January 2019.

My Gurkha team who cleared the compound with us and had my back after the attack in Roshan Village in January 2019.

The DS staff of *Ultimate Hell Week*: Ger, me, OB and Staff at the very beginning of Series One, not knowing how this was going to turn out. (© *Andreas Poveda*)

Dawn on Series Two of *Ultimate Hell Week*. The students are playing in the surf as Alan (right) and I look on, and Rosco films diligently. (*Courtesy of Motive Television*)

A recruitment poster used for ARW selection in 2015, with me climbing out of the Atlantic Ocean onto a gas platform. (*Courtesy of the Army Press Office*)

safely recover the hostages, helped by night-vision goggles, laser targeting and thermal-imaging optics, a major advantage for the mission. The rebels caused mayhem in the area for a few days until the money started to roll again and they got what they wanted once more.

We started to patrol further north and went on long-range vehicle patrols through various districts, each with its own band of thugs or rebels with their own checkpoints. It was all very complicated politically, with areas changing hands daily, although most rebels would run away when they saw us come along with our heavy machine guns mounted on the vehicles.

We did a vehicle patrol to Gbanga (where I'd had my surprise surrender experience) and where a UN Pakistani infantry unit had now set up. On our arrival they greeted us with an orderly who served us glasses of water in our vehicles from a silver tray. They positioned us in an open area of their camp that was vulnerable and often fired at with high-explosive mortar rounds. They weren't setting us up for trouble, though: it was a vulnerable point for the small infantry unit and we were happy to secure the area for them. My platoon boss, a captain and John Wayne-type character, was having none of this mortar business, so we found out where the local head rebel lived and decided to pay him a courtesy visit. Along with my team we pulled up outside this house, all lined up with our heavy weapons pointing at the building.

Six or seven armed guys were lounging on the veranda at the front of the house. My boss walked up to the head man sitting on a chair and told him that if anybody shot at us, we were coming back for everyone, starting with him. Soon after, the rebel leader headed for the hills and went missing. Needless to say, we were never fired on.

◆

The most clandestine operation I took part in was for years only known about by the three men who led it. This top-secret mission, undertaken at night, was to target our own RSM, D, in order to liberate a very specific item of kit. D had an inflatable sofa with an image of Homer Simpson with a beer, which he would lie on in his down time. The RSM was a big man who was tough but great craic and a complete father figure and joker, always setting up lads and taking the piss.

Three of us from my team, camouflaged up with face paint, got our night-vision goggles and infiltrated the small balcony where he kept the sofa. (We were conscious of the high risk of being shot by one of the nut jobs sleeping in HQ with guns by their sides.) We liberated the sofa and moved to the beach where, being divers of course, we swam out into the sea and anchored the inflatable sofa to an ammo can. When daylight came, Homer looked like he was on a lilo in the tropical surroundings, as the surf rolled him gently on the sofa.

Our toilet or 'thunder box' was located right on the beach, with the best view in the world. It was an open-air set up – basically a big metal box that was burnt out daily with petrol. When the RSM arrived for his morning constitution, he spotted his beloved sofa floating on the tide and went mental. He conducted an immediate interrogation of the usual suspects and started slapping a few lads around in fun as well.

Somebody ended up shooting the sofa later, a mercy killing to put the RSM out of his misery. Nobody talked, mainly because nobody knew who it was, and we never told the RSM. I told him 15 years later, when I worked for him in Afghanistan, he couldn't believe it.

◆

There were moments of fun and normality but they were few and far between, and we never forgot the reality of the situation. The week before Christmas we travelled to a small town in the north called Yekepe, situated in an isolated valley that had been a diamond-mining town before the country fell apart. The area was protected by a local Catholic priest who had managed to keep all the rebels at bay using the word of God. We arrived into the valley to the sight of cheering crowds and local people dancing, who were so happy to see us arrive to ensure their safety.

We set up a clinic with our doctor and medics to treat some of the issues the local women and children were having. These people had been alone for many months without any medical help. One morning a man walked up to the door with a tree branch sticking out of the top of his head. He was perfectly coherent and spoke some English. The branch had entered under his jaw and gone up outside his skull, without too much damage by the looks of it. The doc just knocked him out, pulled out the branch like a tree surgeon, gave him some antibiotics when he woke up and sent him on his way.

While we were there one night the priest, using an old projector, showed a movie to the local people on the gable end of his mission building. I sat out of sight in the dark to the side and watched, not the movie but the people as they sat there and cheered and laughed. The sense of community and togetherness was amazing and so evident. The priest had managed to keep them all united with purpose and belief that they would all be safe, all respect due to him.

We left on Christmas Eve 2003 and returned to Monrovia for some down time. We carried out a number of DA missions in the new year, mainly along the lines of snatching the commanders of various militant rebel groups around the country and bringing them back to face justice. The war crimes and savagery between the groups had been horrendous and taking trophies of heads or ears was common. The local practice of 'short sleeve long sleeve' was also in use, where someone's arm would be chopped off at either the wrist or elbow, hence the sleeve-length reference. This was done with a machete as a punishment or warning, part of the psychological warfare used to subdue a village or group of people. The civilian population was terrorised and the breakdown in order just gave these street hooligans a foothold to control the local areas. Women and children were also in the firing line: children were forced into the various groups as child soldiers, while women were routinely tortured and raped. Life was cheap.

We had to return to Yekepe early in the new year as a group of rebels had taken several hostages in the town and were controlling the area, so we couldn't leave those people and the priest at their mercy. They held 30 people hostage in a shipping container and brutalised them, so we went on a mission to clear the town and rescue the hostages, which we managed to do without any hostages being hurt. After clearing the whole area we were picked up by MI-26 helicopters, which can each carry three vehicles and reduced our commute back to base from two days to two hours. We arrived back to applause from the Irish infantry guys, now co-located with us, who had heard about the hostage-rescue mission. We were delighted to have been of assistance to these people who had showed us such kindness. Yet it was strange to be received as such because the rescue had been but a tiny dent in what was occurring daily, with violence and death everywhere. You never get used to the brutality of it but you just get on with it. That was our job.

The north of the country was in the 'blood diamond' mining region, bordering Sierra Leone and Guinea. The south was heavily forested, with several hardwood and rubber plantations dotted around. Some were still operating but most had been destroyed in the fighting, dismantled along with the infrastructure of society.

One day while on patrol we came into a tiny hamlet in the south of the country near the village of Zwedru. Shortly before our arrival a small child had been killed by two men from a rival group, when the boy had stumbled into them at the edge of the village. They had been scoping out the place when they were surprised by the child. They hacked him to death with a machete before escaping.

It's very difficult to witness that kind of brutality on an innocent and it affected the teams deeply, especially those guys with children at home. How do you help people when this type of event is a regular occurrence? Our commander was particularly affected by this event so we went off task for a few days in an attempt to track down these men, but we were never going to find them. They just vanished into the jungle.

I was put in control of a small team in one vehicle, which would push ahead and usually arrived at the little local checkpoints on the dirt roads before the main group. The checkpoints were usually a small hut with a branch or rope across the road, run by the locals, of which faction we wouldn't know until we spoke to them. There was no uniform and government militia looked just like rebel militia – a recipe for disaster as we had to risk ourselves every time.

One morning we arrived very quickly at one such position and came to a halt. I dismounted the vehicle. One man stood in the centre of the road in front of me, I covered him with my weapon as I spoke to him, telling him who we were. He relaxed once he understood we were UN, but just then a second man emerged out of a small hut beside the road. The second man held an AK-47 in one hand, and although he had the weapon pointed towards the ground, he stepped into the middle of the road aggressively, taking in the situation.

I had my light machine gun comfortably across the bonnet of our vehicle, with the safety catch off and the barrel trained on this new guy as he walked out. Mick, in the driver seat, had his pistol on him also and the two guys in the back had their weapons ready.

The man, just 10 feet from me, looked me in the eye. As he took in the gravity of the situation I could see the fear in his eyes. I didn't speak; I was just waiting for him to move the weapon barrel the slightest amount forward, when I would cut him in half with a burst of fire. I already began to take first pressure and squeeze the trigger in anticipation.

He knew what was coming and in a miraculous, split second, decided to take his hand completely off the hand grip. He moved his hand away from his body and the trigger, his left hand on the weapon now in a passive position. As he was no longer an immediate threat, I released the trigger fully forward but kept my finger on it as I remained watching this man. He had been so close to death.

We moved on without incident and I understood again how composure and discipline is the key to all actions. The other men with me could have shot him also, so it proved how the intensity of our granular training puts us in a place where tiny margins decide our behaviour.

Mick and I discussed it after as we drove, with a bit of humour of course, and agreed that our training worked. That man was just protecting his village as best as he could, for all we knew. Yes, it would have been easier to kill him, but our unit goes to great lengths to preserve life and teach you that just because you can kill someone, it doesn't mean you do, until you have to. Then you need to be absolutely ruthless.

———◆———

A critical part of our term in Africa was how we were open to changing our procedures when things didn't work. Everything, from how you carry a pistol when driving to formations and how much ammo you need, is included in this. The lesson is only good if you process it into your organisation and it is communicated to others as a written change of procedure, for everyone to see and understand.

Mission debriefs are important and can be done as an informal 'hot wash', usually completed immediately after an event to capture the main points, when everybody is still a little pumped. The more formal debrief takes place either with all the team personnel or the TLs and main players only with the canvassed input of their team. The second option is usually more efficient. This is then followed up by an after-action review (AAR), which details the operation, what worked and what didn't.

We finished our tour in Liberia and I got some good feedback on my conduct report. More importantly, I was being put in control of small teams. When I returned to Ireland I was made 2I/C of a team, which gave me more direct responsibility. Special Operations is a constant evolution, which will see you move up the structure – but only when you are ready for it.

———◆———

The structure of the unit puts you in a position where you are constantly under a certain amount of stress: here, stress is positive because it keeps you focused and allows you to excel when it is managed correctly.

The 'U-stress' curve is a principle I was familiar with from my PTI training and it is useful in any endeavour. This is the process whereby the higher your stress level goes, the more aware and focused you become, resulting in a maximum level of performance. When the stress level becomes too severe, however, you reduce your efficiency and ability to be effective, ultimately becoming dysfunctional. Eventually, though, you recover.

We needed to establish this scale, both as individuals and leaders, showing us the optimal range for us to operate under stress with the best results. The scale rated from 1–10, with a 5 being the average adult limit of stress. This is the pinnacle of the stress curve or U-stress limit. The average level in Special Operations is above this number. We can, of course, push this boundary and thus improve overall performance, but we need to be aware of the signs of both chronic and acute stress.

Chronic symptoms are usually health-related, such as sleep issues, headaches, illness, loss of appetite and depression, and often remain constant over a period of time. The immediate physical signs of stress include palpitations (thumping heart), nausea, chest pain and breathing difficulties, all common reactions as the body pumps adrenaline and other hormones in preparation for the upcoming challenge. The acute symptoms for an immediate stress manifest as more psychological and can include aggression and irritability, emotional instability and reckless behaviour. I had a guy in my team in East Timor experience this: he had a complete breakdown and had to be committed to hospital while on leave, where he remained for months.

———◆———

I carried on with various things for the next months of unit adventure until I started as a student on a survival instructors' course run by unit instructors. That's what I loved about the unit, there was always something new to learn. It would qualify me to teach people how to survive off the land with nothing except maybe a penknife and a smile. After weeks of instruction you have a test period where you have to survive off the land (in my case somewhere in the West of Ireland) for a week, building shelter and fire and trapping and hunting game. It gives you a good idea on how you have to adapt to keep yourself alive.

In real-life survival situations people often don't survive – they just give up, overcome with isolation and loneliness. The survival training process trains self-belief and a will to succeed, as often the only thing that stands in your way is yourself. The course culminated in a period of becoming the hunted, as you were chased like a fugitive for as long as it took to get you. After this you step into the psychological underworld of what is called 'conduct after capture' or resistance to interrogation. You become the prisoner and after sleep deprivation, interrogation and constant harassment you are brought to the edge of sanity. Your grasp on time and reality becomes unstable. I learnt to cope, as we all did, with nobody being immune to this type of training after a while. The take-away was a belief in my strength of mind and my confidence in getting something completed when I committed to it.

My next educational experience took place when I was sent to the UK on a light recce commanders' course (LRCC) – or how to manage and run reconnaissance operations in all phases of warfare; basically gathering information. My friend W and I were the only two international students on the otherwise British Army-only course, which would be run in the reconnaissance school in England and in the Welsh mountains. The Britishness of the course was brought home to me when, on arrival, we were handed a large manila envelope containing course instructions stencilled with 'At Her Majesty's Service' in huge red lettering. *Here we go*, I thought to myself, *my republican grandfather would love this.*

The infantry-based course was run at a frenetic pace and involved various fitness tests and battle runs to gauge ability, along with a written test on British Army doctrine. I was put into a team of six, two officers and four NCOs, made up of paras, a marine and a Gurkha. We lost a guy on week two when he broke his leg, so we carried on with five, spending the next two months crawling, patrolling, digging and falling our way around Salisbury Plain and the Welsh mountains, with large amounts of kit on our backs. The grim winter weather emphasised the importance of preparation, good leadership, discipline, attention to detail, effective observation and personal administration. The motto of recce is 'The timely and accurate delivery of information,' and emphasises the importance of discovering information and getting it to the people who needed it. Accurate information wins wars.

Communication was crucial. My team would change appointments every two to three days, with each man leading or taking other positions and responsibilities. I was constantly delivering or preparing formal orders, to be given to a group in minute detail, with a scathing debrief by instructors if they were off. If you are expecting people to follow you and carry out your plan at all costs, then you'd better be very clear on what you want them to do.

A good procedure to follow is Who-What-Where-When-Why: **Who** is carrying out the task, **What** is the task, **Where** is the task, **When** does the task need to be carried out and **Why** does it need to be carried out. Communication is the key. 'No comms, no war', as the saying goes. If you can't communicate clearly with your team or command structure, it's over.

I had an instructor for my team, a colour sergeant from the Parachute Pathfinder regiment (PF) who didn't have much time for the Irishman on his course. He had it in for me from the start. My teammates, all British soldiers, rallied and supported me so when I was in command they would bust their asses to ensure success. The instructor eventually softened his attitude, especially once we were on exercise in the mountain, where he could see my skill and capabilities. Eventually he even had some grudging respect for me.

I had a funny experience where we were calling in live artillery fire onto targets from a bunker system nearby. Our job was to observe the impact and radio in corrections as required for accurate fire. The artillery instructor running the shoot was putting us under pressure to deliver accurate details to the guns, located some distance behind us. We all took turns on the radio and each time he heard my lovely Irish lilt he would start with the 'C'mon Paddy' and 'Let's have ya Mick,' as he stood behind me.

At one point I got so fed up with this that I said to him, 'Give me a break here, Colour – I'm used to lining up gas bottles from the back of trucks at your barracks, not this fancy stuff' (a reference to an IRA method of firing bombs onto British bases in Northern Ireland at the time).

There was complete silence from the 20 British soldiers around me in the enclosed bunker. I was thinking, *Shit … too much?* when all of them exploded into laughter, the instructor loudest of all as he tapped me on the shoulder, saying, 'Fair play, mate.' Army humour is army humour regardless of the army.

My team of five and I spent weeks alone, sharing everything and helping each other to get through. You know someone is a friend for life when he's holding the freezer bag you're going to the toilet in. Our observation posts (OPs) were dug underground, close enough to a target so you can observe all movement without being seen. We had to carry everything in and everything out after the operation, leaving no sign, so that means bodily waste also. Another lesson learnt: ensure you can tell your urine bottle from your drinking-water bottle. An easy mistake to make in the dark, so remember never to drink from the warm one …

We often had a few beers on the weekend and had a great course blow-out at the end, coinciding with the Black Watch (Scots Regiment) who were just back from Iraq and out on the town in Warminster, resulting in an inevitable platoon punch-up. It's worth mentioning that when this regiment was on the town, the local police force had riot police and horses on the streets, like you would see at a soccer game.

W and I did very well on the course, finishing up as two of the strongest candidates. It confirmed to me once again that attitude, effort and commitment will get you a long way, regardless of the size of your army. The British Army is a huge, war-fighting machine with hundreds of years of tradition and experience, whereas the Irish Army is a small force deployed in a different manner and can't compete on a larger scale. Yet when you break the regiments down into small groups and teams, you find that both armies operate in similar ways.

The time in the UK also confirmed to me that in order to be a functional leader there are four points to bear in mind:

- ▶ You need to be able to plan and prepare well.

- ▶ You need to have personal leadership skills.

- ▶ You need to be able to execute your plan correctly.

- ▶ You need to know how to learn and adopt lessons.

———◆———

I returned to Ireland just before Christmas 2005 and spent it at home with Sinéad, which was very nice for a change. I was now being lined up for a four-month sergeants' course in the NCO Training School of the Military College, which would see me back to the regular army so I could qualify and be promoted to TL in the unit.

The course included a mixture of senior corporals from all over the army, everything from cooks to clerks, tank drivers to air stewardesses. I learnt some good lessons, the main one being how many really good people are in the army. There were also several strong female candidates and three in particular were very impressive. The attitude and effort by one air crew member was outstanding; she put most of the males to shame. She was brilliant. She had all the qualities that are important in a good soldier and was sound as well. I know better than to underestimate the power of females, growing up under the reign of four big sisters.

Unfortunately a certain element of trainees, both male and female, were incredibly negative and just wanted to do the bare minimum to pass. Some would moan about everything: 'Why are we doing this test? ... That dinner was shite ... I'm hungry ... I'm wet ...'

There is a saying in the army that 'Even Jesus had a crib,' but nothing undermines the fabric of a team more than someone constantly pointing out the negatives for the sake of it. Look, we all have a little moan now and then – I'm not naive, nor am I Mary Poppins where I sugar-coat everything, but I am a realist and even when I acknowledge the negatives, I balance them against purpose and meaning. If you want to succeed you have to concentrate on the positives (while of course being aware of the negatives) yet understand how you can mitigate those negatives – or just suck it up and get on with it. My first CS had a great saying about people who complained: 'You signed the dotted line. And all you're entitled to is 18 inches on the square' – a reference to the manual of foot drill where you stand at ease, the regulation stating your feet are no more than 18 inches apart on the parade square.

———◆———

I finished the sergeants' course and was glad to get back to the unit that summer of 2006, where a number of us would be interviewed to fill two TL positions. My divorce – which was just a formality – came through the day I did my interview before a board of the CO, RSM and another senior officer. I felt nothing, only a sad sense of time wasted, really. I was by now long over the emotion. The minute I stepped out of the interview room, I was straight into a car and off again to the naval dive school for a combat dive supervisors' course, to qualify as a diving operations instructor. It was intense, this time more so for the management: we had to organise dives, diagnose and deal with dive emergencies and injuries as well as several other tactical elements, all in all 95 per cent thinking and 5 per cent action.

I was home again by Christmas 2006. The day we broke up we ran a 10k race and then went on the beer, as is the tradition. During the festivities the CO rang me to tell me that I was going to get one of the TL slots and my own team. All I wanted to do was be a soldier and to make sergeant like my dad and grandfather before me, so I was over the moon. A good friend, John – known by everyone as 'Killer'– was also promoted alongside me. It was a moment he had been anticipating for a long time as he had been in the unit for several years at this stage. He wouldn't drink that night since we had a medical next morning for the promotion (he had been close before and lost it for various reasons) but I drank my ass off.

I was living the dream at home – so much so that Sinéad and I went off to the US on a holiday of a lifetime, the highlight of which was getting married there. I'm old school – we were settling down and talking of having a family, so I wanted to be married before we did that. We had a tiny ceremony with two witnesses and a wedding video that you might actually watch because it's less than seven minutes long, including two songs!

◆

In the spring of 2007 I was lined up to run the unit skill course in Ireland, in which myself and a team of instructors would be teaching new guys, just off selection, the full package of basic skills. It's an important job, as you are responsible for building a Special Forces operator from scratch. The students, a mixed bag of rank and experience, are stripped to the bone and rebuilt to suit the skills required for the activities of the unit. Officers, NCOs and privates are moulded into a cohesive team – although, of course, not everyone makes the cut.

I had a syllabus to follow but could use my initiative to make the training exciting and testing, and this I relished. I was a hard but fair taskmaster. I worked closely with the group, observing skills and flaws alike in a microscopic manner, with the final say on who passes and who fails. If someone is not effective in training they become a liability on a real operation, so you can't take any chances.

The responsibility of deciding who makes it and who doesn't comes with the job, so you have to be straight. I never balked from cutting guys, even several months into the process and God knows how much time in preparation for them. Some mistakes are red-line issues and warrant instant failure – usually weapons-handling skills, integrity or suchlike – when students are a potential danger to themselves or others. They need to be able to fire live ammunition within millimetres of friendlies, with no margin for error.

The students are constantly being put under pressure, so they learn how to problem-solve and deal with adversity as a matter of course. This results in self-reliance and resilience, leading to an inherent self-belief in what they can achieve.

In my leadership role I also learnt how to be cognisant of the other instructors on my team, to bring on their qualities as leaders and teachers and to empower them with the autonomy to take ownership of aspects of the course. You must be confident enough as a leader to comfortably allow others to take control as is required; you don't always have to be the voice, the authority or the one who comes up with the idea. The trick is to understand how humility works and be aware when you are important and more importantly when you are not – 'surrender your ego', as the Queen song goes.

When I delegated a task I expected it to be completed with diligence. If things went wrong and it was a genuine mistake we would resolve it as best we could and move on. This is psychological safety in practice.

Psychological safety is a high-performance-culture system of understanding that things go wrong and mistakes happen. The person who makes the mistake has to know that if that mistake was genuine and a result of human error, there is no reproach or punishment. This gives them and the group the confidence to keep doing their best, thus significantly improving overall performance. If someone did not give a task enough due diligence or respect then I had no sympathy. I'd come down hard and give less latitude in the future.

In one case I ran into an instructor in a ragged and dirty combat uniform as he left a formal lesson with the students in a lecture hall. I pulled him aside and quietly gave him a good talking to, reminding him that as an instructor he had to be the perfect example at all times, that sloppy dress is fine in the field but not in the classroom. It may seem like a small issue, but if you want to create a high-performance culture, you can't take shortcuts.

CHAPTER 7

ECHO TEAM

was starting to take on more projects, having moved from being the first man in the door to being the guy who sends the first man in the door. I was thinking more than doing, organising large numbers of activities and people at the same time. I am not a natural multi-tasker but as a leader I was now having to deal with more issues created and managed by others directly or indirectly under my command.

Working in Special Operations is a juggling act between the individual expression of yourself and your skills and the giving of yourself for a group goal. There are four things I learnt early on about how to manage a lot of assets, people and operations at the one time:

▶ Know your limits: personnel, assets, resources and most importantly your own. You have to be absolutely realistic about what you are trying to achieve, the time you have to do it in and your team's ability to do it. This will allow you to concentrate on what you can control and then not to worry about what you can't. Getting this right it saves a lot of trouble down the line.

▶ Learn how to prioritise activities and actions into what is critical and must be done immediately, what is important and needs to be done quickly and what is routine and can be pushed out.

▶ Properly delegate tasks and projects to others and fit in how you supervise this. Some people you can trust to complete tasks without supervision; others need a bit of eyeball. It's a fine line to navigate, since micromanagers just piss everyone off.

▶ Plan ahead when you have the opportunity to do so. This will give you the best chance of success. Plan your own time before you plan your team's. The army rule of thumb for planning is to allocate one third of available time for you to plan your part, with the rest of the time allocated to your team's contribution and preparation.

The course finished in that autumn of 2007 with some outstanding soldiers of all ranks in the group. I was thrilled to set up an intricate final exercise for them, culminating in them arriving on a site and being presented with their green berets by the CO. I didn't do it on the Hill of Allen, as was customary. I really wanted it to be a shock and surprise, which it was: 'Always expect the unexpected.' It almost feels like being a parent at an offspring's graduation when you stand beside each man as he is handed that green beret and puts it on, the sheer joy and sense of achievement in their faces as you shake their hands.

———◆———

I moved back to my day job as a team leader of Echo dive team in the maritime task unit. We conducted our team training and operations and came together with other teams in training blocks at various stages of the year in order to keep on top of current skills and techniques. Skills fade in diving is a problem, so these training blocks are designed to ensure that everyone is brought up to speed on the latest techniques. Regardless of rank or position you can find yourself fitting into any training slot of a dive team to get slick in your drills once again.

Part of my remit involved conducting dive training on a gas platform off the south coast, which we would 'attack' in a counter-terrorism role. One of these scenarios involved a covert attack with a climb from the sea onto the platform to clear it, although we practised other methods of insertion also.

During one such training event I was part of a team on an attack dive, with the plan of climbing up a ladder system on a main support leg of the gas rig itself. Myself and two of my team were halfway up one leg, a 15-metre climb to the first deck. The steel rungs of the ladder are bent and broken from the heavy swells you get 30 miles out, so the climb is not easy. Waves come up and completely submerge you as you climb, so you have to take a breath and just hold on, continuing up when they recede.

As we climbed, one of my team called a warning – a huge, freak wave was coming in from my right. I braced myself against the support leg and held on as hard as I could as I looked out onto a wall of water that was now blocking out the sky as it broke well over our heads. There was no holding on with that bad boy; we were all washed away as the wave crashed down on us. I went sub-surface for what seemed like ages, tumbling around underwater, like being in a washing machine. We all popped up eventually with no injuries but a lot of coughing and spluttering and made our way back to the leg and reset for our climb and continued on with the exercise, as you do. We estimated the wave to have been 5–8 metres high, coming from nowhere in a relatively calm sea. It was just the gods telling us that on the grand scale of things we were pretty insignificant – always a good reminder to get, now and then.

———◆———

Being in Special Forces isn't just dangerous on operations, where the bad-guy risk is usually mitigated by speed, surprise and violence of action – training can also be life-threatening. People think we are these high-octane snake-eaters fuelled on adrenaline and death-defying thrills – well, some are I guess, but not all. Risk is always there on operations and you are constantly having to deal with the variables. We never gamble in a situation, and will always stack the odds in our favour to ensure success in a live operation. Risk assessment is part of our preparation both for real missions and in training. The risk on a real job can sometimes be incredibly high, but is assessed and accepted by all as part of the mission so we deal and live with it. In training we have a detailed assessment process where a training risk is identified and numbered on a scale of 1 to 10 – it is training after all, and not real.

However, risk is still part of the job and although we push the envelope in training, we have safety measures and procedures with which we mitigate the danger. But the nature of the beast means that from time to time people will get hurt, sometimes seriously. And like all organisations we have a written process to follow for risk assessment. Risk management, however, is what you have to do with that risk as well as how it fits into your plan and affects decisions on operations, eliminating, transferring, mitigating or accepting risk of death or injury. This is my general plan of execution:

- ▶ **Avoid**. Change your strategy to avoid the risk completely.
- ▶ **Mitigate**. Take action to reduce the risk (protective equipment, medical support, weapons, etc.).
- ▶ **Transfer**. Get someone else to deal with the risk – then it's no longer your worry.
- ▶ **Accept**. Decide to take the risk. All military actions and plans involve a degree of risk.
- ▶ **Share**. Spread the risk across multiple teams or support units.
- ▶ **Contingency**. Always have a back-up plan for potential risk and further things going wrong in your execution.
- ▶ **Enhance**. Bring more people to the fight and enhance your team with additional forces or supports.
- ▶ **Exploit**. Taking the opportunity of speed, surprise and overwhelming force or a quick change of focus or direction as it's presented can counter a risk.

Special Forces operators are also what is called a 'force multiplier', which means they punch above their weight or numbers because they are better trained, better armed, better equipped and more focused in a fight – the trump card in the army deck, if you like. We are also in an occupation where the application of lethal force is part of the mind-set, and this deals with a lot of the risk management.

I have been injured a number of times in training: broken bones, concussions from blasts. The worst was a back injury that had me on morphine for weeks and sleeping sitting up on my sofa, as I couldn't lie down. It took me ten months to fully recover and get back with the teams and a huge effort of resilience, training, discipline, patience and diligence to get there. It was tough on Sinéad also, who had to put up with my cranky ass at home.

My outlook changed again when my son was born at the beginning of 2008. It is a wonderful thing to have this little man in our lives. I kept my mouth shut in the delivery room – nobody in there is interested in your opinion – and the experience confirmed to me again how amazing women are. Having a child definitely made me reassess my priorities and, although nothing changed at work, it reiterated the importance of thinking before jumping in.

◆

I was back with my dive team in early 2008 on a training dive in the naval base where we were conducting covert sub-surface boarding drills on a naval vessel tied up alongside the quay. It's a strange sensation, being under the keel of a ship in the darkness of night. You have to do everything by feel – it's not a comfortable place to be in, and panic will set in if you lose focus.

In military or combat diving the idea is to get to the target underwater without being seen, so for that we use oxygen rebreathers, which basically recycle the gas you breathe so there are no tell-tale bubbles to the surface, unlike in scuba diving. The benefit is that the duration of the dive can be increased; but the danger is that oxygen can become toxic under pressure if you go too deep or if the dive is too long. The ultimate result of this is oxygen poisoning of the central nervous system, which can very quickly cause multiple problems and even death.

The dives were short ones, as we were concentrating on the tactical side of getting onto the ship. We had conducted three or four attack swims on the target vessel in controlled conditions and were getting ready to leave the surface to perform another evolution. I was at the rear of the formation beside my dive buddy, C. We were swimming along at the correct speed and depth in poor visibility when I noticed C was starting to go a little deeper and being practically beside him, I pointed at my wrist-mounted depth gauge to get him to correct his position and depth – normally, staying above 10 metres is your safety margin.

I got no reply in the form of the normal thumbs-up hand signal to confirm my direction so I looked along his silhouette and saw he was kicking his legs very slowly. I grabbed his demand valve (DV – the mouthpiece you breathe from, which is strapped behind your head) and pulled it to get his attention. There was no reaction so I grabbed his chin and pulled his face towards mine. The visibility was so bad that I could only see when I stuck my face right up to his. His eyes were completely rolled back in his head, which meant he was only semiconscious. I had to react very quickly as I suspected it to be oxygen poisoning, also known as an O_2 hit.

The pair in front of us were just out of sight – I could only see the flash of a black dive fin every now and then. I sent a quick signal along the line to alert the rest of the team and proceeded to take control of C as I unclipped both of us from the device we used to keep the team together.

I was just about to inflate his buoyancy control device (BCD), which would bring both him and me to the surface, when he suddenly took a violent seizure. As a dive supervisor I recognised what I thought was an O_2 hit, meaning I had to hold him at depth until the seizure subsided as he arched his back, convulsing uncontrollably. I had to stay completely composed and focused and keep him in control as best as I could. I was also very aware that if I got this wrong I would kill him. If I brought him up during a seizure his airway could close and I would give him another dive injury, or even kill the two of us on ascent.

I waited for the seizure to subside, just as M, another diver on our team, came back to help out. M assessed the situation and I quickly gave him the hand signal to go for the surface. He controlled the air valve while I maintained a physical hold of C and vented off excess air from his BCD so we could control our ascent.

We managed to ascend at the correct rate and broke the surface. M immediately attracted the safety boat, which is always nearby on dive training operations. I inflated the BCD fully to keep C out of the water, and immediately undid the strap and pulled out his mouthpiece to get him off oxygen and breathe ambient air, which quickly relieves the symptoms of an O_2 hit.

As he gained consciousness he immediately started shouting and fighting us – this is typical of someone suffering an O_2 hit where the diver is coming out of an almost comatose state and returning to a fully conscious one. The nervous system is still feeling the effect of the O_2 hit so disorientation, resistance and aggression are not uncommon in this phase. The boat came alongside and recovered him and got him straight to the dive medical officer where he was stabilised, making a complete recovery. Had it been a night dive, our normal tactic, I wouldn't have seen him in the water and he would have died.

The positive for us was that the constant practice of drills and procedures and a detailed plan and brief prior to every dive operation will ensure that you are equipped to deal with a situation as soon as it goes wrong. You just have to train for the worst possible scenarios. 'Train hard, fight easy,' as we say in the forces.

———◆———

I was heading overseas once again, back to Africa but this time to Chad, a former French colony on which the French still had a huge influence. The problem area was the Darfur region, which straddled neighbouring Sudan and Chad and had several different rebel organisations and groups fighting over it. The area was barren and mainly desert and I would operate as part of the EU-led operation to bring stability to the region.

I would be attached as a CS to an intelligence team, which was part of the Joint Special Operations set up made up of Special Forces units from France, Austria, Belgium, Sweden and Ireland. It was a bit like Eurovision with machine guns. I was part of a smaller international team involved in the collection and processing of information, a follow-on from my specialty in surveillance and reconnaissance (SnR).

On the ground I spent a great deal of time piecing together parts of information. I also learnt a lot about the analytic side of intelligence, as well as interacting with various SF Task Units gathering the information, which was a lot more fun. Chad, despite its violent history, was a beautiful place, especially to the north of the country where we moved into the Sahara desert, up towards the Libyan border. The area was full of snakes and camel spiders, which were a monster of a thing attracted to light, so nobody slept on the ground at night. If you flicked on a torch you could actually hear them scurry towards you across the sand.

I spent four months in the dry heat of the desert, but it wasn't as humid as West Africa or the jungle. The odd bit of rain that did come would sweep in from the open desert from the north. It was torrential during the wet season, allowing plants and bush to grow so that in places the desert became quite green. The rain would often be followed by a violent sandstorm, which would regularly take away our tents and equipment. Yet I found them fascinating, the way they came up out of nowhere. It was such a rush being in one.

I picked up a bad bout of food poisoning. It came on extremely quickly but luckily I was in our SOF HQ, a large military base located beside a large town called Abeche, where I was able to stagger to the Italian field hospital from my tent. I pushed in the tent door and collapsed on the floor in front of the orderly. I woke up three days later with the headache from hell and feeling like a pin cushion with all the drips in me.

I had to stay in hospital for a week or so and get pumped with fluids, but was busted out by a French commando team I worked with one Saturday morning to watch a rugby match on satellite TV with them. They opened the side panel of the tent behind my bed and pulled me out, intravenous drip and all, and loaded me on a waiting jeep and off to SOF HQ for the match. The Italian medics scoured the camp looking for me, but I remained hidden until full time. On my return to hospital, I received the best (and most enjoyable) 'bottling' (telling off) I ever got from the matron, all in Italian, with passion and hand gestures like someone directing traffic.

———◆———

I bonded very well with all the teams, particularly the French guys. I had good basic French but learnt a lot more from the guys who schooled me in military French – lots of exotic swearing. I also learnt to sing 'La Marseillaise', which always went down a storm when we had a beer, making me a sort of honorary Frenchman.

The French guys are given a *nom de guerre* (or alias) when they make it into special forces. This name they have as long as they serve. I made lifelong friends with a guy called Stephane, who I lived with for four months in the desert and got to know very well. When he later visited me with his family in Ireland, I learnt his real name was in fact Laurent.

Along with a Belgian colleague we found a bar in a backstreet of Abeche, which was a sprawling, mud-walled town full of settled nomadic tribespeople on the edge of the desert. The place looked like a movie set with single-storey houses and camels wandering around unattended. The bar (using the word very generously) was rumoured to have Guinness – well, I was working in intelligence so if I didn't investigate and confirm the information, I wouldn't be doing my job. We found it, a desert speakeasy with hardy people in it, and yes it had crates of bottled Guinness, which were brewed in West Africa and brought here by camel or something. What a place to find it! So we got some and dished it out to the rest of our grateful team on Saturday night. It wasn't great, but beggars and choosers, right? It tasted good enough that you could pretend that you were actually drinking Guinness, which was all that mattered.

In September the Irish Minister of Defence at the time, Willie O'Dea, was visiting the region and all the Irish locations along with the chief of staff, Lieutenant General Dermot Earley. Myself and M, an officer with me in SOF HQ, waited on the desert airstrip for his aircraft, a French Military C-160, to come to a halt. Once it did, the ramp came down slowly and as the dust swirled around him, standing there was the chief, who stepped off the aircraft and immediately and deliberately strode towards us with a warm smile and a bearhug of a handshake. I had met him on a number of occasions: he was the absolute poster-boy image of what a leader should look like; a naturally commanding man in crisp fatigues and a smartly positioned beret, comfortably chatting to us in the intense heat.

The minister unfortunately didn't cut quite as chiselled a figure as he disembarked unsteadily from the aircraft, pale-faced, the jacket of his crumpled suit thrown over his shoulder. He wore black, dust-covered dress shoes and a white, open-neck shirt –

not ideal in the 40-degree heat – and paid us scant attention as he was greeted by senior French officers. (I had met the minister some years previously during an ARW display in Ireland when a colleague and I were demonstrating weapons we use in operations. He had picked up a Sig pistol, which got him into hot water in the press for posing with the pistol. After that, he tended to keep his distance from us.) Looking the part is half the battle the army says, and I learnt the truth of that after this episode.

I made some great friends in Chad and learnt a lot about working and collaborating with different nationalities, languages, politics and outlooks. It is easy to work with different people if you take the time to understand what they are about and always try and find the good; the spark of something you can work with. You don't always have to like the people you work with, regardless of their nationality, but you do have to respect and get along with them to achieve the mission.

---◆---

On my return to Ireland after Chad I spent a few weeks getting to know my new son who had been four months old when I left and was now close to nine months. I'm sure he was wondering how did this guy get out of the computer screen, because that's how he saw me every chance I got while away, on Skype. I loved playing with him and he got used to me being around, with the three of us now a little family unit. But too soon I was back into the maritime task unit as a TL once again. I was now very much developing my own leadership traits, not just using the ones I had copied from others along the way. We all follow our own settings in the end!

We had now a core group of senior TLs that set the tone for everything else in the group. We were also pushing the boundaries on our capabilities and developing new skills and techniques, almost as pioneers of our craft, under the guidance of the senior NCO – an outstanding, forward-looking individual with great foresight and also a very good friend.

I was involved in a number of training developments with other Special Operations units around Europe, completing some short-term training attachments. The process involved a team of us being embedded in their unit while training on how to conduct various counter-terrorist activities. This was a great experience – the sharing of ideas, tactics and outlook showed me that we were on the right track, and for a small unit we were absolutely punching above our weight in ability, experience and equipment. So much so in fact, that some of the more high-profile units sought us out to play after they saw a video of us assaulting a Stena Line ferry in the Irish Sea. The evolution would see us board a huge passenger ferry en route to Dublin by hitting it at sea while it sped along at 24 knots. The advanced procedure is incredibly technical and extremely dangerous, as we climb from small, fast boats up the side of the rolling ship as it slices through the rough swell and waves. The penalties of a mistake on these assaults are the most severe if you fall. Our skill at handling the fast boats in these conditions and our ability to covertly board a moving ferry pricked a few ears and eyes – I guess they didn't expect a little country like ours to have this ability. As a UK maritime SF unit commander told us, 'Nobody has the monopoly on skill or experience.'

———◆———

After months of deployment around the world and little time spent on Irish soil, it was good to be back home for a spell. I was delighted to find out that I was to be the senior sergeant instructor on the next selection course. Although a great honour it is also a huge task, involving months of preparation prior to the course itself, which began on a Friday night and would be one of the last to follow the old syllabus of four weeks of selection followed by six months of basic skills. The numbers of students who turned up to attempt the course was low, for reasons unknown – maybe nobody wanted to be a Ranger that year. I remember standing beside the Unit CO who had been the course officer on my own selection course while we watched the young DS doing their thing. I remarked that we might have a problem, with which he concurred, but we had to carry on regardless.

The course, designed to test core values, stress and fears, was extremely tough, and saw us lose most of the candidates by the end of the first weekend, despite them being in the top percentage of ability and accustomed to always doing well. In their home unit they are probably rarely under pressure and when they are, they operate at 75 per cent. On selection they operate at 100 per cent all the time, and some just can't handle that they get things wrong or fail on tasks. This will often get into their heads and make good people quit as a result – just as I did, the first time around.

By week two we were down to just three candidates, and then there was one – who would have two of the most intense weeks of his life to look forward to, after which he would spend months in training. It was incredibly hard for this one man, a young officer, to keep himself going without anyone to share the load. He cut a lonely figure, going everywhere on his own with a DS running beside him. The ratio of DS to students was now 15/1 and some of my instructors were beginning to feel sorry for the young officer, but I had to put the boot in: the student simply had to continue as was required for the duration of the course. We even issued him with a buddy – in this case an Action Man figure, complete with all his kit – so at least he had someone to talk to, just like Wilson, Tom Hanks's volleyball friend in *Castaway*. We're nice people like that.

I planned to carry on with the course regardless but a crisis meeting was called with the unit's senior management, myself, the selection course officer and the CO. The CO canvassed us on our opinions as to our course of action. Everyone except for me and the unit deputy commander said we should cancel the course.

I was furious. I said to them, 'If we axe the course now, what message are we sending out? This man is giving his all to get through it. The least we can do is give the same back.' The deputy commander, a good friend who had served in Chad with me, agreed and backed me up, so the CO had a good think and told us to come up with a contingency plan, which we did.

We reshuffled everything and I put a young unit member in with the candidate to give him a real buddy of sorts – Action Man had quit. Of course, it would have been much easier to cancel the course, but I was bound by loyalty to that student and the system to complete this task. He struggled on for several days and after a particularly hard 14-hour night mountain march, carrying a 90-pound backpack, he broke down. I could see him begin to unravel. He walked over to me, tears in his eyes, and said, 'DS, I want to RTU.'

I looked at him, remembering when I had been in his shoes on my first selection course. I said, 'OK, I want you to take a minute and think about this, then come back to me.' He stepped away and turned back almost immediately, saying, 'Nah DS, I'm gone.' 'Fair enough,' I said. I shook his hand and got him out of there, back to a shower and a meal.

I made a point of seeing him before he left, I told him how strong he was, that in my eyes the mountain didn't beat him, and that he would have passed the course had he not been alone for so long. I asked him to go away and have a good think and come back on selection again because he had the core values we wanted. He thanked me for our team professionalism and my words before he left to go back to the army. He came back on selection soon after and passed well, going on to become the unit deputy commander later. A positive word to give someone something to work with can never be undervalued. It had worked for me.

Despite everything I felt the course had been a failure. The assets, time, personnel and effort we put in for months had nothing to show for it. I was bitterly disappointed as I would have liked to have had some students who passed, just to experience that moment when they finish.

The course photograph was taken and placed beside the framed photographs of all the courses that had come before it, as is the tradition. What is not the tradition is to have a course photograph completely bereft of students with only DS staff in it. It was the only time in modern times, to my knowledge, that no one passed a selection course. I remember an ex-Ranger soon after congratulating me for being a gate-keeper and holding the line for standards, and although I took his comment in the spirit it was meant, I was disappointed. There would be no new blood that year, and if you can't keep recruiting good people you won't survive.

◆

I had some great news at home when we found out that Sinéad was expecting for the second time, and were very excited to be getting ready to increase our own home team. I had gone to the early scans with her and all had been well, with Sinéad feeling good and strong. But over Easter Sinéad had a bleed. We immediately went to the hospital and got the scan to make sure everything was OK. I was terrified like never before, sick with worry, feeling absolutely helpless while uselessly trying to comfort Sinéad. They couldn't find a heartbeat. I watched the medical staff desperately trying to locate it, but as the time goes on you see in their faces that it's gone. The baby did not survive, the heart just stopped beating.

I tried to make sense of what was happening. How could this happen to us? I was usually so good at dealing with things and could compartmentalise well, but this was different. When we arrived home from the hospital I hugged my son as hard as I could, silently thanking whatever power allowed him to be alive as I tried to construct a course of action. I was composed and controlled but I had lost something inside that I will never replace, so I just concentrated on Sinéad and trying to help her. The first pregnancy and baby had been perfect; I thought this would be the same. You really only understand what can go wrong when the worst possible thing happens. The innocence of that part of life is gone. We never think it will happen to us, naively assuming that those things only happen to other people.

On Easter Sunday I sat on a chair in the hallway outside the operating theatre as Sinéad underwent surgery to remove the baby that hadn't survived. I had been so impressed by her strength during the birth of our first son, and now I was absolutely moved by how she dealt with this tragedy – with decorum, resilience, strength and selfless love for all of us throughout.

Sinéad got pregnant again the following year and it was a double-edged sword for me because I was so happy in one way but terrified throughout the whole pregnancy. She was calm and took it as it came, although for me it's probably the most worried I have been in my life. Our daughter was born, perfect and healthy, later that year. I shed tears of joy when this beautiful little person arrived in our lives, I was that happy and relieved. Children totally mess up your mind-set and soften you up, the little feckers.

◆

I spent an extended period at home, getting to grips with the new woman in my life, despite getting about as much sleep as I had been getting during selection. At the same time I was replaying the course in my head and thinking of changes to how we recruit and the process we run. I was going back to the unit soon and was getting ready for some more intensity, but with a difference – I would be doing specific bodyguard training.

After some time off with my family, and putting the failure of the selection course behind me, I moved on. Next up on the agenda was something completely different – bodyguard training. The beauty, I always found about this type of work, is that you have to be ready for absolutely anything at all times and be completely selfless in how you operate. Everything you do is for the smooth running of the mission and safety of your principal (the key person you are protecting), most of which your principal won't even notice. This has the beautiful effect of paring my not-insubstantial ego down to nothing because your only job is to stay in the background, speak only when spoken to and protect your principal.

We protected mainly the highest-level government officials, ministers and others from the Department of Foreign Affairs and Defence as they travelled in countries that warranted a protection team. Like any other operation the devil is in the detail and it relied on getting on the ground ahead of time to recce locations. Some principals loved having a protection team; others were embarrassed and self-conscious. The main part of protection operations is protocol, which takes about 75 per cent of your time on duty. Protocol, in protection-speak, relates to the things that enhance the principal's prestige, like them not being late for an appointment or leaving out the correct report on the back seat of their car for them. It is also important that they look like a swan gliding along and nobody sees the furious kicking below water. The remaining quarter of your time is spent on security, protection and risk, for which you rely on accurate, timely and constantly updated information. Bodyguard training runs at various times to keep up skills or when a protection team is actually required either for short trips abroad or for months or years on a rotation basis. After I had completed a period of time of training my next assignment would be to go back to the regular army for a while.

---◆---

In the first months of 2010 I was attached to the NCO training wing of the infantry school in the Military College, to be an instructor on an army sergeants' course. We would regularly attach instructors to various schools to keep up with training methods as well as to input and share techniques from SF – at least those that we could divulge.

I was responsible for the Infantry Tactical and Patrolling phase of the course as well as running the physical training. The CO was a good but tough guy who wanted me to be the bad cop. He was also a ball-breaker, who wanted his course run along supremely intense lines.

The first so-called tutorial exercise was a three-day tactical event, which should have been a straightforward tutorial in the field, a walk-and-talk-through, but ended up being nothing less than an ordeal. Three guys from my unit got no more than six hours' sleep during the three days, which is fine if you want to torture people, but not good if you are trying to instruct them. At the end of the three-day period the CO ordered the group of 50 students to sit on their packs while he walked around them debriefing them on the exercise as we stood and looked on: the guys were done and barely awake, never mind taking in any new information. When he finished, he handed over to me for the next part of the debrief.

I called the student orderly sergeant and said, 'Right, get your platoon up and get them to bed. We'll do this fresh tomorrow.' They came alive in an instant, grabbed their packs and disappeared, much to the disgust of the CO. He came at me right away, hissing, 'What the fuck are you doing, Sergeant Goggins?'

I stood my ground and explained that you can't learn if you don't get enough sleep. Students need to be relatively fresh and rested to absorb information. He was looking at this as a chance to show them how tough it should be. I was looking at this as them learning these skills properly so that one day they too would be able to lead platoons of young people on real overseas missions. The army mind-set was a constant problem, in that things were taught and completed for passing a test on a course and not for real-world activity. You need to be crawling on your belt buckle, experiencing hardship and pressure on a regular basis to stay current and connected to what really matters. In the end we came to a gentleman's agreement and he backed off and let me get on with it. As I said, he was a good man.

———◆———

When I finished in the NCO school and on my return to the unit I was promoted to senior sergeant to manage the complete maritime task unit. I was delighted. I was putting my stamp on the teams, I thought, but a few weeks in I noticed that there was dysfunction between the team guys and all the sergeants. I investigated the issue and discovered I had been complicit in an elitist, exclusive culture. I was part of this for a while without really knowing, both as a TL and now as the senior rank. I had to take a good look at myself and the sergeants – all my good friends – and review my leadership. After some hard conversations we managed as a group to rejuvenate our bond completely from top to bottom, mainly by including the younger and junior leaders in more crucial decision-making and development. Once more we had to instil a sense of ownership and a buy-in mind-set to all our personnel, which we did in time, creating a better culture. We didn't make it a republic – I was still running the show and would bark when it was required. (I heard Jack Charlton once said that to be in charge you have to be a dictator, but be a nice one.)

I relearnt that you always have to keep your link with your people. I was confident in my skin and took this issue as a good thing, not a challenge to my leadership, as I might have in the past.

CHAPTER 8

FROM HERE TO TIMBUKTU

We settled into a good culture and routine as a task unit (TU) and I was also delighted to see some of the younger guys step up and take on leadership roles. It wasn't all high fives and 99 ice creams, but I was always happy if we made a realistic and committed effort to our tasks. I didn't rule with an iron fist, more like a Kevlar one, where I gave fellas a chance but if someone was taking the piss I would nail him. If someone disrespected what we were about or took short cuts, thinking themselves more important than the group, I was hard on it.

Where possible I would deal with this myself and mete out punishment or 'corrective action' as it's now called. I usually listened to a guy's side of the story before I acted, people can do some pretty stupid things and swift action is often the only countermeasure. I never recorded anything officially if I could help it, preferring to iron out transgressions with a good 'bottling', with extra work, or with a 'dirty detail' – a shit job to fit the crime.

One guy got himself into some serious trouble outside of the unit and did a really stupid thing. I pulled him in and told him that I no longer trusted him and that he was absolutely useless to me and I wanted him to go. I was very hard, and could see he was devastated by my words.

Well, he didn't quit and knuckled down and in time became an excellent asset to the TU as a sniper. Incidentally, I found that good snipers have to be a little more self-focused, which is fine for a sniper – not so much for combat diver. The lesson for me was that people can change and surprise you. I had the guy written off and he proved me wrong, so I told him that. Every day is a school day and we just keep learning about ourselves and other people as we go, adapting our view all the time.

◆

I was doing a lot of recreational distance swimming at the time, which I had gotten into after I had injured my back. Having complete twisted two facet joints in my lower spine, I couldn't run for a long time so got back in the water. I thoroughly enjoyed it, entered a number of civilian events, and continued to swim in the sea, lakes and rivers with a group of guys from the unit when we could.

During the summer of 2012 I was at a friend's house for a summer weekend near Lough Derg and while both families went for a walk one morning, I went for a long, solo swim in the lake. I began to swim quickly and warm up to go the two kilometres to a small island and back to the little pier I'd set off from. With every stroke, each time my face was in the water I could see the black depth of the lake below me. It was a still and quiet morning, with a light mist, which reduced visibility. For some reason I swam directly out from shore towards the island instead of my normal routine of swimming along the shore, which is safer.

After an uncomfortable ten minutes where I couldn't get into my stride, I could feel a slight cramp, which I pushed through, but minutes later my leg completely locked out and I stopped dead in the water. I couldn't kick. I was wearing a light neoprene bodysuit that would normally keep me afloat in sea water if I stopped, but this was fresh water and I was less buoyant, so I began to sink. My head went under the water a number of times as I gasped and struggled with my leg, thinking, 'I'm going to die in a lake in Tipperary.' The shame of it! I looked around – it was like being inside a cloud now, the mist so heavy, and I couldn't see beyond 20 metres. I wasn't sure of the direction or where I had come from, which was probably no more than 600 metres away.

I started to do a rescue side stroke in the direction I judged was towards the shore. I kicked hard with my strong leg, which began to cramp up as well. With neither leg working I was sinking every few strokes and I could feel a panic starting to rise up as my arms were becoming tired from the effort of working to keep me afloat. I realised that if I carried on, I would become completely exhausted and quickly drown.

I work in water and am usually comfortable in it, so I made myself get a grip, closed my eyes and took a deep breath and went into a survival technique. I stopped all movement, tilted my head forward and allowed myself to sink feet first under the water. The air in my lungs naturally stopped my descent after two to three metres, and being positively buoyant slowly brought me back to the surface. Once I felt my head break the surface I immediately exhaled and inhaled, before sinking again and repeating the process. We call it 'bobbing for air', a combat water-survival technique I learnt to drown-proof myself.

The cycle takes 15 to 20 seconds each time you sink until you surface again for a breath, so you get three to four breaths per minute. The secret is to let your body go and not to move at all. I continued this for around ten minutes, saving energy, focusing and reminding myself why I needed to make it to shore. I was getting cold now, but I felt strong enough to try and swim again, so I did. I listened out for sounds from the shore and heard the noise of a vehicle in the mist for a second, then side-stroked in the direction it came from. I moved slowly for several minutes until for a second the mist cleared enough that I saw some bulrushes 100 metres to my right. I went for them hoping I could stand on the roots under water, which luckily I was able to do. *Ya flute*, I thought to myself.

With my mouth just out of the water, enough to get a breath, I swam through the bulrushes, standing where possible. I could hear the noise of my children on the bank. I hit a sunken fence line and followed it in to where I could finally stand, then walked out of the water to my family and friends as if nothing had happened.

It was a stupid, overconfident thing I did, not giving myself a safety margin. I was lucky I had been trained in water survival, which saved my life. I had failed to show respect for the water so it gave me a slap to remind me. It could easily have been the end of me.

———◆———

In November 2012 I was invited back to Collins Barracks in Cork along with many others for a ceremonial parade, as a guest and former member of the 4th Infantry Battalion. The battalion that my family had served in for generations was to be 'stood down' or scrapped in some political sleight-of-hand move by the

Department of Defence as a budget cut or some other nonsense. The current soldiers would be absorbed into other units and an artillery unit would take over as the senior garrison unit in the barracks. I stood in a doorway near the parade square in civilian attire, watching as the battalion colours and its members, current and former, filed past a reviewing stand with some token general taking the salute before the colours were laid down for ever. The unit served the state since its founding with its members giving their lives to the country and it was discarded by a minister of defence who did what was convenient for votes to suit political power instead of doing what was right. It was a very sad day for the army and for me personally, with my heart still very much a part of the battalion in which I had so many friends and in which someone from my family had served continuously since 1951.

When I was a boy growing up in Cork, we would refer to Timbuktu as a place so far away it might as well be on the other side of the world. 'He can go to Timbuktu as far as I care,' people might say. I always wondered what and where it actually was but I never imagined that I would be anywhere near it in my lifetime.

Mali was experiencing a huge insurgency problem that had swept the Sahel region of Africa, which divides the Sahara desert in the north from the humid savannas to the south. There were constant attacks in the north of the country by IS, Boko Haram, Al-Qaeda and other supported terrorist elements around Timbuktu. An EU mission was planned: it would be broken into a number of parts, with Ireland sending a small detachment as a training team.

It was an historic event in that for the first time since independence, we Irish soldiers would be under British command. We were required to travel to the UK to join up with our British unit for training, an element from the 1st Battalion Royal Irish regiment. The Irish team was made up of five army NCOs, including myself, and a young lieutenant on his first overseas deployment who, as it happened, was the son of my old FCA training sergeant, Eddie.

We arrived at the British base on a Sunday night and were organised quickly and billeted in the Sergeants' Mess, where we were met by a senior NCO who gave us a welcome drink of sorts. We quickly gathered he wasn't a fan as he was obnoxious and proceeded to slag us off about being Irish Army. In fact, he was a complete gobshite. He said the RSM would meet us the next morning and that he would make short work of us. I wanted to flatten him, but I remained composed. All in all, it was not a good start.

The next morning we paraded on the square for the RSM, who is a God-like figure in the British Army. As he made his way down the line looking sternly at the Irish team, he stopped dead in front of me: the RSM was none other than Spence! We had been students on the recce commanders course together years before. 'Well Ray, what about ye?' he said in his thick Belfast accent as he grabbed my hand and pulled me into him. He immediately arranged for a get-together to welcome us properly and, true to his word, that night we had a great catch-up and lots of pints.

◆

We landed in Bamako, the capital of land-locked Mali, in early April 2013 where, along with the UK team, we numbered around 30 and would be commanded by a UK major for the duration of the mission. We loaded our weapons and kit and drove in a tour bus like tourists to our new home, Koulikoro, a small urban area 60 km from the capital located on the banks of the great Niger river. The town was dominated by the vast river and stretched along the main road with a soccer pitch of sorts every 100 metres or so. Mali means hippo – the river apparently being full of them – although I never saw any, even when I swam in it later. Koulikoro was a Malian army-officer training barracks, and it was here that we would receive a company of Malian soldiers to train for the next ten weeks.

The UK guys were all young with little or no instructor training experience and, being an Irish regiment of the British Army, the group was made up of 90 per cent Irishmen from all over the island of Ireland. Of the total number of us in the group of 30, just four were actually English and one of them was in our Irish Army team. Gerry had been born in Nottingham (but was more Irish than any man) and then there were three 'proper' English men. We also had a Fijian, a Scot, a South African and a Pacific Islander, showing how broad recruitment was for the British Army.

It was decided that the Irish contingent would be broken into three teams and mixed with the UK guys, mainly because we were more experienced instructors and also to keep an eye on us in case of a rebellion, I suppose. I was to be the deputy commander of a team led by a brand-new British subaltern, just out of Sandhurst, who spoke like a prince. It soon became obvious that he was out of his league here so the major made him company training officer and me commander of my team of five UK NCOs. We worked well together, and they had some fantastic skills that I encouraged and used to the best in training.

While we were preparing for the first induction of Malian soldiers I thought hard about how I would get them to accept me as a leader. *What do people fight for?* I asked myself. *More importantly, who do people fight for?*

Firstly, people fight for people they respect or like, someone they believe in. This I could manage. Secondly, they fight for someone who will bring them success even if they don't like or respect them. This I could also get. Finally, they will fight for someone they fear, but only for a limited amount of time because eventually they will not work or fight for fear. My strategy was to get them to fear letting me down, so training was going to involve a combination of all of the above.

When they arrived that Friday we had an officer, a platoon sergeant and three groups of about 12 men, each led by a sergeant and supported by a corporal. They were processed by our admin team and documented, with passport photos taken of each man. We would begin training the following Monday morning at 0600, so I spent the rest of the weekend memorising faces, names and backgrounds. It wasn't easy: the names weren't exactly Smith or Jones so I had my work cut out, but preparation and attention to detail will get you off the mark.

I was completely aware that these men were coming to us to be trained to make themselves better to protect their country from insurgents. They were willing to travel the huge distances from their own areas and form contingents to keep their families and people safe. So it was up to me to give them my best shot as a trainer. Malians are very respect-driven and paying compliments is a key part of their military structure, with constant saluting and parade-square movements the norm. When I first approached our platoon of troops, the Malian platoon commander immediately shouted *'Gardez-vous!'* bringing his platoon to attention. He then saluted me, even though he outranked me. I told him to give the command of *'Repose!'* or 'at ease', then gave them a simple introduction in French as to why I was here. I told them that if they gave me their best effort, I would make them an elite fighting force. What won them over was my next gesture: I stepped up to each man individually, greeted him with a handshake and called him by his name. They were surprised and delighted by the attention and my effort. It goes to show the importance of taking the trouble to get to know people as individuals, because the more respect and understanding you show towards each other, the better you can work and live together.

Nevertheless, there was always the possibility that some of these men were insurgent terrorists who could take out instructors, known to us as an 'insider threat'. To that end, we always had at least two armed instructors, with one teaching and the other covering covertly in case of an incident. We also searched the platoon before every evolution and anything we recovered, mainly bullets and grenades, we would give to the commander to be taken back to their camp. They had just arrived from a combat zone prior to getting to us. After a few days we found less and less ammo and soon there was none, until we handed it out for them to fire on the shooting range with us. We still searched the kit before every lesson as a safety measure and to ensure that live ammunition is not accidentally brought into a training environment. Safety being a huge part of weapons training in any army, we worked initially on safe weapons handling with our platoon. We had already mitigated against them shooting us intentionally – we now had to ensure they didn't do it accidentally.

I also had another problem to deal with in that the platoon, although all Muslim men, was made up of two main ethnicities. Most men were Bambara, sub-Saharan Africans who made up the majority in the country. I also had a smaller group of Tamasheq men, originally nomadic tribesmen from the Sahara Desert. Both groups spoke different languages, with French the common tongue, which meant interpreters for French, Bambara and Tamasheq. So not complicated at all during lessons. These two groups had fought each other for centuries. The Tamasheq people had in the past captured Bambara people and shipped them down the Niger to become slaves. In addition, many of the current Islamic insurgents came from the Tamasheq region. Fighting broke out on a regular basis in the camp and between training events, with the odd stabbing also.

How was I going to unite these men and stop them from killing each other? Very simply, I had to be an effective leader. A leader is someone who unites a group of people and manages them in a common purpose. I had to use all my EQ (emotional intelligence) skills to bring them down to basics and get them to work together.

I began by making buddy pairs and creating situations where they had to depend on each other. I placed the Tamasheq lads into each of the teams, while also instilling a sense of responsibility in the junior leaders to look after *all* the men, not just their own.

Besides training them in all the basics of soldiering we also had to coach and mentor the leadership group in how to motivate and manage the troops. Sometimes they would arrive for classes with men missing, and when I asked where they were the Malian NCOs would say, 'Oh maybe they are sleeping.' There was a lot of that. I would then go to the tents and hunt down the missing soldiers, dragging the Malian NCOs with me to show them that this was their job.

NCOs disciplined their men with physical violence and bullying. As much as I abhorred it, I had to tread carefully and not interfere, which I did by encouraging the NCOs to administer push-ups instead of a beating. Push-ups became almost a unit of currency, thrown around for all kinds of mistakes and issues. Even the instructors and myself did push-ups when we got it wrong, which the Malian lads loved.

Our time in Mali coincided with the dry season, with temperatures rising to more than 50°C some days. The barren scrubland was covered in small, loose rocks, which absolutely ate up boots and energy and made the long 48-hour patrols a workout.

We worked from Monday to Saturday, 0600 to 1800 in the training camp, with a 48-hour period spent patrolling and living off site once a week. Every evening I did my own physical training, then prepared lessons and ideas for the next day with my team. Lights out was early. As I always say, sleep is a weapon: if you get enough it's your weapon, if you don't it becomes someone else's. Sundays were a rest day, when we would catch up on some personal admin and just relax and plan the following week. A lookout point on a hill inside the perimeter was my favoured location, where I could sit on a rock in the evenings and observe the fishermen on small dugout canoes on the river, just 200 metres away. First chance I got I was going in it. Despite the hippos and health issues the only warning I was told of was not to urinate in the river as a small parasite in the water would swim up the urine stream into your urethra as you went – being barbed, it would be painful to get out, I'm sure.

I had some run-ins with our UK leaders over various issues. The British Army is probably the best infantry army in the world, but boy do they know it. Humility is not high on the list, so they don't always listen to other ideas or opinions. The major and colour sergeant had their own plan, which didn't always fit here, although to be fair we sometimes agreed or they eventually took advice. I respected both men because they were all for the protection and welfare of their men.

Speaking of welfare, I received a clothing resupply from the British Army, sent from HQ in the UK. It was mainly socks and T-shirts, but also contained some British Army underpants. I never saw that coming. I also received a care package from an organisation in the UK where people make up a shoe box of items and send them to a soldier serving somewhere. It was a tradition that went back to the Great War. My package came from an 11-year-old boy in Yorkshire and the letter inside began with, 'Dear Soldier, I hope you are well.' I was glad of the puzzle book and the biscuits and he also sent a drawing of his home. I think it's an amazing and timeless tradition.

I bonded with the Malians over physical activity, not only in carrying out military drills but also in playing recreational sports. Soccer, rounders, touch rugby – anything I could think of to create a common ground that was fun and stopped them wanting to kill each other. Getting them to compete as teams against each other and gelling that team as one was my mission.

They started to get it. In the mornings we would run as a block or platoon and I loved the loud, proud songs the men sang as we ran around the camp. I noticed a soldier one morning wearing combat boots with no soles. I could see the skin of his bare foot as he lifted his feet to run, but he wasn't going to stop. He completed the run without complaint so I didn't act then, but I couldn't have this so we managed to get him a pair of boots (with soles) later that day. The men called me 'Go-Gins' – no rank, no other label, just that – and shouted it in unison during their songs. They were starting to believe in what we were doing and, more importantly, believing in one another.

Slowly but surely we made progress. I was able to bend them to what I wanted from the group on a basic level and started to push further into the tactics employed in a counter-insurgency war. I had some excellent UK instructors and our lessons were always interesting in that they were field-based and teaching took place outside. There was no death by PowerPoint as our classroom was mostly under a tree, using a chalk board or bottle tops to demonstrate tactics. Several of the platoon could barely read or write and the majority had little or no formal military training of any sort. We concentrated on discipline, teamwork, loyalty, rules of war, tactics, weapons-handling and medical training, with just ten weeks to deliver a fighting platoon.

One problem we ran into was teaching the difference between cover from fire and cover from view, so we devised an exercise where we got the platoon to stand on a small hill, filling their helmets with small rocks and waiting for my order to throw them. We would pick out groups of five men and get them to approach the hill tactically, using the ground and trying to stalk and stay out of sight as much as possible. I then ordered the 35 men on the hill to throw as many rocks as possible at the stalking men in an attempt to simulate bullets coming at them and to demonstrate what will stop a bullet and what won't. It was a good evolution and they enjoyed it immensely, plus I got my message across about using cover – or so I thought.

The Malians wore a tribal charm around their necks called a 'gris-gris', which is a small pouch of material containing various items that ward off evil spirits and also conveniently stops bullets. They believed that in battle the gris-gris would keep them safe and stop them from being hit. When I asked them why they'd been hit by rocks on the exercise despite wearing their gris-gris, they replied that it only works with bullets! Fair enough, I said. I carried my own keepsake, wooden rosary beads given to me by my mother when I first went overseas. They had been my father's, given to him as he boarded a truck to go to the Congo in 1960. Neither of us was particularly religious, but both of us carried it on all our overseas missions and came back in one piece, so maybe there is something in it.

At the end of the ten-week period we held a parade and graduation ceremony for the platoons and companies. The Malian platoon had committed to what we were doing and were open to everything we tried to show them, and that's all you can hope for. My platoon gave me two gifts by way of thanks: the first was a tribal chief's ceremonial medallion, presented with great ceremony from the platoon commander. I was very proud to receive it. The other gift was my very own gris-gris, because they said they didn't want Go-Gins to get shot. I also didn't want Go-Gins to get shot, and graciously accepted the gift.

◆

We returned to our own training, which included planning future courses and analysing how the previous training had gone. This would take a couple of weeks, after which we were due to fly home for two weeks of leave, which I was looking forward to enormously. We had a bit of down time also and were allowed to go to an old hotel nearby under escort to sit by a pool and have a beer. It was right on the river's edge and I seized my chance.

I paid a fisherman for a loan of his canoe so a colleague and I paddled out a bit and took turns dropping covertly into the water. It was out of bounds and forbidden so we were being bad boys but it was great to finally feel the cool water of the river around me, and getting out of the heat for a minute was heaven. I had missed being in water so much, as it always gives me such a level of well-being; I didn't stay in for long though as I was paranoid about the hippos I never saw and the little creatures that swam up into your mickey if you peed into the water.

We carried on with preparation and counted down to leave, which was only a couple of weeks away. One night I received a call from Sinéad: my mam had had a stroke and was barely hanging on to life. I needed to get home immediately. The Irish lieutenant colonel in charge of training was slow to react to help me – in fact he did nothing. It was my lieutenant, John, who arranged the flights. The problem now was getting to the airport in Bamako. It was hours away and I couldn't exactly call for an Uber.

The British Army stepped up to the mark. I remember the major saying, 'Right, Sergeant Goggins, we are getting you on that flight if I have to drive you myself,' which in essence he did. He squared away an escort team made up of Irish and British soldiers and three vehicles with him physically leading the operation and personally ensuring that I made it home. Although we didn't always see eye to eye, he was a man of honour and absolute integrity who I respected greatly. He put himself on the line for his men, me included, and I will never forget his effort for me.

◆

During that long journey home, all I hoped was that I would make it. I just wanted to see my mam one last time and remind her what she was to me and how she inspired me in life. As I travelled, I thought about all the things she had done for me and for others, her outlook in life, how she fought for her family and selflessly gave herself anytime she could.

The memories came flooding back and as I sat in the plane looking at the world passing by. I realised that my mother, in fact, had all the qualities of a good soldier and leader: all the qualities I am highlighting in this book. The greatest soldier in my family wasn't either of my grandfathers with their war medals, or my dad, or my brother or myself. It was my mam.

To me a soldier is someone who gives themselves selflessly to others with purpose, integrity and commitment, for their protection and to save others, which she did every day of her life. She was also a crack shot – well, with a slipper anyway.

I made it to the hospital in Cork some 32 hours after leaving Koulikoro. All my siblings were there in a vigil around her, as was our way of dealing with things, together, the way we had been taught. My sister Catherine said, 'Go on into her, she's waiting for ya,' just before I went in. I entered the room alone and sat with her on the bed for a short time, talking to her about what I had been up to in Africa: Mali, the great River Niger and Timbuktu.

She wasn't conscious but I'm sure she could hear me as I reminded her about how she had carried me to hospital in her arms as a child, how we had listened to the choir in Notre Dame Cathedral and how she had held my newborn son with tears in her eyes. A force of nature herself and a great woman to talk, it was strange that I would never hear her voice again, a voice that always brought comfort. I always just wanted to make her proud as I was cutting through my life and as I sat there holding her hand for what would be the last time, it struck me where a large part of my own courage and resilience came from.

My mam died peacefully the next day after a life of love and devotion to her family and doing good things for people, never thinking of herself and always protecting us no matter how big we thought we were. If that's not a leader then I don't know what is. She was even selfless in her last hours, I like to think, holding on for me to come home, because she knew it would have broken my heart not to see her one last time.

She had a fire in her that made her live life on her own terms, to keep fighting no matter what. The person who left the most indelible mark on my life at all stages was now no longer here but yet will always be with me. We buried her with my dad, to reunite them again, as there was always only one life in the pair of them.

◆

I had some time at home, and having come back early for my mam I now rolled into my actual leave and went on a short holiday with Sinéad and the children – it was magical and I appreciated every second with them.

I returned to Mali soon after to finish my deployment, travelling back with the training team who had all been at home on leave. We flew to Lisbon where we met up with the UK team and continued the rest of the journey on a Tap Air flight, where I alone had been bumped up to business class, much to the disgust of the Irish lieutenant colonel, who had to sit in economy with the other mortals. Sometimes you just have to enjoy life's little twists and turns when they happen.

I continued to train another platoon of Malian soldiers for another ten-week period following an improved set up with us more experienced in what we had to do. The company we trained previously was already in combat in the north of the country, with reports that they had fought well but returned to type and were carrying out operations along their own guidelines again. We had spent a lot of time teaching them about the rules of war and some civilian UN teachers had gone through several lectures on human rights and the rules for non-combatants. We had our work cut out in rules of engagement.

The rainy season began, sending torrential downpours and intense humidity. I also experienced the loudest thunder I ever heard in my life, with an amazing deep crash and rumble that shook the buildings like artillery fire. The Malian platoon I now had was not equipped for the rain and nervous about operating in an environment that turned into raging streams and mud in minutes. The landscape transformed overnight and what had once been rock and dust was now covered in bush. It changed completely how we had to patrol and the tactics we could teach.

Like the previous platoon, they were a good group and we trained them to the best of our ability. One guy was bitten by a tiny snake and was near to death almost immediately. Luckily he was medevaced out with our South African air ambulance team, which saved his life. The snake, a viper, was smaller than a 12-inch ruler with incredibly toxic venom – the medical team had never dealt with this particular type of venom before. Like the rain brings out the worms at home, it brings out the super death snakes in Mali.

———◆———

We completed the Mali mission in the autumn of 2013 and our composite training team was replaced by another Anglo-Irish team. The main instructors travelled back to Britain in a military transport, with the platoon leaders and management to follow.

We returned to the UK on a Royal Flight executive aircraft, which was a small, spiffy plane, ideal for the ten of us to fly in luxury. The short-hop aircraft would need to refuel in both Dakar and Spain en route to London. The flight time was endless. Two of my colleagues decided they would get a sedative from our German medical colleagues so they would sleep all the way. I was dubious and desisted, which turned out to be a good call. The two boys, instead of falling asleep, were actually in a stupor, which saw lots of drooling and terrors as they hallucinated all the way to London. It was incredibly funny and the best in-flight entertainment I have ever had.

◆

After a short period of leave I was back with my TU and into diving operations as soon as possible. I continued with various domestic operations and training until the early summer of 2014 when myself and a colleague, Dave, flew to Tampa in Florida for a training exercise and attachment to US Special Operations Command (USSOCOM). The purpose of the training revolved around a huge urban Special Operations and maritime counter-terrorist workshop and training exercise. Dave and I would be part of an International Special Operations Force (ISOF) event, run by the US, which would host hundreds of operators from all over the world who would create small teams and conduct a large-scale training event in downtown Tampa.

We stayed off base in a hotel in the city, which housed most of the maritime SF guys from all over, numbering in the hundreds. Strangely, the same hotel was also hosting the National Santa Association of the USA, so when we arrived I noticed Santa sleighs in the car park, along with a pen of reindeer and various yuletide-emblazoned vans and pick-up trucks.

Inside the hotel, in the restaurants, bars and the pool, mingled some 400 professional Santas and Mrs Clauses, who were there for their annual convention … in May. They were all completely in character at all times, even donning Santa swimwear in the pool. It was one of the craziest things I have ever been involved in – 400 Santas and an equal number of Special Ops guys who were on a bit of a jolly and partying every night after training completed.

Training culminated in a public demonstration of capabilities where the mayor of Tampa was a hostage rescued by us as we powered into the city-centre harbour in US Navy jet boats. The whole event was for show and the area was surrounded by media and thousands of spectators who were reacting to each manoeuvre like a crowd would at a football or hurling match. I found it all quite surreal. It wasn't something I was used to and I kept my face covered to the cameras. The US lads in our teams, SEALs and Marines, weren't too worried by it at all, especially when girls were writing their phone numbers on tennis balls and throwing them into the exercise area. It was definitely a good experience and a bit of a busman's holiday to be part of a team where every second guy was a different nationality with different opinions, different attitudes and different-sized egos.

———◆———

Back in Ireland I jumped straight into preparing a combat diver course I would be leading. The administration is pretty intense because I first had to get all the potential divers medically tested by the naval dive medical officer before they could even get wet. Once they pass the medical, we then have to run dive aptitude tests to check if they can even function underwater and that they have no issues with breathing and pressure changes. This is all done safely in a dive pool situation with instructors and medics on hand because there is usually someone with an underlying condition he knows nothing about that the water will always pick up.

We began the ARW phase of the course with a multitude of lectures on dive medicine, physics, seamanship and dive tables and signals. I worked them very hard for weeks – probably too hard in some cases, as two very good guys blew their eardrums.

The next part of the course was tactical, involving long, covert swims to hit targets using newly acquired navigation skills. I got the impression that they were slacking off, so I decided to impose a sense of urgency. Towards the end of the course I pretended to be really angry about a comment a naval instructor had made about them. In the middle of the night I gave the divers a quick-dress order and got them to board the dive boat fully dressed and armed, telling them on the way that I would make them earn their dive badges tonight. I could see the big eyes in each dive mask, full of apprehension as to what was next.

Just as I was about to put them over the side of the boat in the pitch dark of a winter's night, I got one of them to open a medical box, which was part of our boat kit. It was full of bottles of beer, which we had stowed away in there earlier. Each man, much relieved, got a beer, and I made a toast of 'Here's to us and those like us!' They had all but passed the course in my eyes and it was a good way to finish it with just a written test to come, which I knew they would ace.

This would be my last major input with the maritime unit. A few weeks later I went to the CO with a two-year suggested training plan for the TU. He looked at it, told me it was great and then announced that I would be going into the training detachment permanently. I understood his reasoning, but for me the whole idea of serving in a Special Forces unit was to be in the teams, not a training school. I needed to think seriously about my options and where I wanted to be – maybe, even, to move on completely and leave the unit for good.

CHAPTER 9

CHRISTMAS PARTY IN KABUL

S pecial Forces operators are usually capable people who are in control of their surroundings at all times, or so you would think. A saying I heard once goes something like: 'The two rules of Special Forces are, always look good and always know where you are. If you don't know where you are, then just look good!'

Even in the thick of a situation you have to stay composed and project absolute confidence, whether or not the arse is falling out of it. Never let your adversary see your weakness, but if help is available you need to able to reach for it also.

In some cases, guys will not look for help with a task because they feel it will be seen as a failure. I have witnessed guys close to collapse trying to complete something that they will never get done alone before they reach out for help – or have it forced on them. We all want to look competent, but a good leader will realise that there always comes a point when you need help.

Three vitally important things show how secure someone is as a leader:

- ▶ Be the first to admit you are wrong. It shows complete confidence in yourself.

- ▶ Ask for help when you need it, regardless of the situation.

- ▶ Learn to read the room and understand the mind-set of the people you are dealing with.

Let me give you an example of reading the room, I was part of a large-scale skills and equipment demonstration in Army HQ in Dublin for a US chief of staff, which had basically stalls set up from all kinds of groups like bomb disposal, Navy divers and many others. Each group gave the VIPs a detailed rundown on their capabilities. The group of generals inspecting the stalls then broke for lunch and on their return, started the rounds again. My stall was set up towards the end. I had a unit special-reconnaissance vehicle fully kitted out and a detailed brief to go with it, but I could feel the vibe from this group and I could also smell the brandy they had had at lunch, so I wasn't going to waste my breath and their time. I opened up by simply saying, 'Gentlemen this is a vehicle we use in reconnaissance operations. Any questions?' I could feel the instant relief from the group. The Irish chief of staff even came up and shook my hand as he thanked me for a quick and painless brief. Ultimately, I was glad that I had taken the time to read the room, even if it meant curtailing my detailed speech.

———◆———

I was sent to a leadership seminar with other NCOs to discuss experiences. The army teaches a very formal type of leadership, and from my experience as a leader in Special Operations and in the army, I can break down my style of leadership into two approaches.

▶ The first is a consultative approach, in which a team leader takes advice from his team in relation to a decision, which the leader will then make on the basis of the information received. This works well when you have time, but not in a quickfire, critical situation.

▶ The second is the dictatorial approach, where you give people simple but definite directions to complete an immediate task. When people are in the thick of it they don't need choices or options, just direction.

People also need to be able to recognise moments when they can stop and calm themselves or prepare for an immediate event by using tactical breathing – this is a composure measure, designed to settle soldiers just before a stressful action. We have been doing tactical breathing for years, but now it was being taught more deliberately. The process is simple: you inhale through the nose on a count of two seconds and exhale through the mouth on a count of four seconds, for two or more cycles. The process will calm and refocus an individual and bring them back to a steady, composed state.

———◆———

At that time, Sinéad's father was very ill and in hospital after a heart attack. I remember going to visit him with the children. Sinéad went down to get him a coffee with the children while I sat on the edge of the bed chatting to him about sport and history, both subjects he loved. We enjoyed a good, simple relationship – I suppose I never needed to be involved in any more depth as Sinéad was always around. During the chat he suddenly stopped and grabbed my hand and held it, looked at me intensely and said, 'Ray, this is the best time of your life now, you have to enjoy it.' It was a very profound statement and I took it to heart. He knew he was dying, which he did shortly after. It reminded me that you need to appreciate what you have, try and enjoy what you do and be happy on a daily basis. 'Happiness is not the destination, it is the journey', I once heard someone say.

Our home life was also happy. My son was well settled in school and my daughter kept us on our toes, the little devil. It shows the spirit in people, even at a young age. But soon heartache came knocking once again, when Sinéad, who was newly pregnant, experienced a bleed, which meant we lost this little soul before they even had a chance. It's hard to lose someone when they have lived. It's a tragedy when they haven't.

———◆———

By now I was a fully fledged desk commando, neck deep in drafting a manual to comprehensively incorporate some of the training we had developed and used in the unit. It was a monster task and took me several months to piece together. It would, in fact, be my swan song in the ARW. Here was I hoping for a blaze of glory; instead I got Operation Microsoft Word as my final mission. This task, however, showed me that I longed to stay active and that I still had a bit of time to try something else before I was too old.

My good friend Killer, who'd been promoted to sergeant at the same time as me and with whom I'd served in the ARW, was diagnosed with terminal lung cancer just as his wife found out she was pregnant with her first child. Along with some of my good friends we set up a chemotherapy rota, organised by Baz, and we managed to have a good laugh with him – typical soldiers' black humour. Despite having the craic, I could see him fade over time, but the kind of resilient man he is, he was going to fight it and see his little girl being born, which he did.

———◆———

I applied for my 'ticket', as we called discharge. It was a huge decision to leave a place that was my world, in which I had grown up and that had given me so much for so long. It was in my blood and a big part of who and what I was, where I had met some of the best people I would ever work with, some of the best mentors and teachers, but mostly some of my closest friends.

Two reasons convinced me to leave. The first was to start something new while I could. The second was that Sinéad and I were struggling financially. The ARW pay had been cut to a ridiculous level in the downturn and I'd had enough of being underpaid for the hours put in, not to mention the risk. We were living on one wage, as Sinéad stayed at home once our son was born. I never expected to make a fortune in the job but I now needed to provide for my young family, and that made up my mind.

I went on pre-discharge leave on a Wednesday in September 2015 and walked out the gate as a unit member for the last time. Instantly, I became someone else. The nature of the beast is that guys move on and adapt very quickly. You are out of the gang in a heartbeat.

Two days later I flew to Paris to start a civilian job I had lined up – CP of a huge French rugby player who was twice my size and probably didn't need a bodyguard. I was going to be part of a protection team that would work for MasterCard and on their behalf for the 2015 Rugby World Cup in London.

I worked on a consultancy basis through a third-party company, reporting directly to a Donegal man and ex-soldier named Gavin. We hit it off from the start. He was a smart guy and showed me the system of security and protection in the corporate world, which is similar to what I had done in the military but not as formal. As a rugby fan myself I had to escort groups and VIPs to all the matches over two months, as well as setting up the security team and events under Gavin and another ex-Ranger. This was the man who had brought me into the job, having been a mentor of mine since I started in Special Forces.

I escorted a large group to the final in Twickenham, which ended long weeks of events for the team and myself. As a rugby supporter it was a dream come true to be at the World Cup Final, sitting in the stand, working, yes, but still there. The other memory was of being off the clock and having a meeting in the foyer of a hotel in central London. Jonah Lomu, the famous All Black, and his family were part of our VIP group and vaguely knew who we were. He saw us and came straight over with a friendly 'Hey fellas,' then had a bit of craic with Gavin and myself. Neither of us are celebrity-conscious, we were just chatting away with him as normal, which I think he enjoyed. A charming, honest and gentle man, who would, within weeks, tragically die as a result of a kidney condition.

For the rest of that year and into most of 2016 I went on to work with MasterCard in various high-profile protection tasks with the CEO and other senior board members. The company has a huge involvement in sponsorship of events like the Brit Awards, the Champions League and several other public events. I was very interested in the leadership traits of the top executives and was lucky enough to protect Ajay Banga, CEO of MasterCard, and Vice Chair Ann Cairns on occasion when they travelled in Europe and Eurasia. They were two amazing people, the epitome of what leaders should be in their interactions with people and the example they set. They had a kind word for everyone, from the doorman in the hotel to the girl who serves lunch. I watched them as they conversed at the tempo of the people they were with, whether CEO or bus driver. It was an education to see the light touch with which they managed people, confirming once again that being a good leader hinges on how you create relationships and deal with others.

I was in Paris with Ajay Banga for the Roland-Garros tennis tournament. As part of the advance team, myself and other team members would go ahead to check locations before the VIP's arrival. I was at the match venue very early in the morning, checking the routes in and seating locations to ensure a smooth entry later that day. I was sitting in the MasterCard box at 0700, finalising the plan for the day, when I became aware of two guys playing tennis. Although I'm not a tennis fan, when I heard grunting I looked up and noticed one of the men looked familiar.

It was none other than John McEnroe, playing a training game with Pat Cash. I watched for over ten minutes as these two giants of the game played with skill and intensity, with me the only spectator in the enormous stadium. It made a huge impact on me, to witness the grace and elegance of these athletes and towering figures in the game.

Working in that environment I learnt so much about how to develop the more subtle skills of VIP protection and tact. I'm sure I was still quite rough around the edges, as I had spent my life being aggressive, decisive and acting with assertion. I was learning now that kicking in the door isn't always the way – sometimes you need to open it gently.

I was involved in the MasterCard 'Active Shooter' programme, which is a workshop for staff on how to deal with a terrorist attack, or life-threatening situation in the office or elsewhere, something MasterCard took very seriously. We ran this programme in a number of locations in Europe but the one that sticks in my head the most is the one we did in the Paris office just two days after the attack on the Bataclan theatre on 13 November 2015, where 130 people were massacred by an active shooter. The staff were frantic as there had been several follow-up attacks in France at the time and our task was to educate people on how best to react to such an event. We explained the precautions to take to avoid such a situation as well as what to do if you found yourself in one.

The main point is to be aware of your surroundings and quickly assess a location when you get to it. Most terrorist situations can be avoided if you have an 'eyes and ears up' outlook as you move around in larger urban areas, being present and aware of what is around you – and not pretending the rest of the world doesn't exist, like most city dwellers. If you are in a building, locate yourself where possible so that you are not near windows or other entry points, or on balconies or sitting outside on the street. Identify an area in which to take cover quickly if needed and check out an escape route for yourself. If you do these three simple things you are five seconds ahead of everyone else.

Check what is happening in the areas you may travel through in advance, and sign up for security updates from your embassy or department of foreign affairs if going abroad. You don't have to be paranoid but in the world we live in now it's just sensible to take precautions. If an attack does come, the principles are simple: run–hide–fight, in that order. If you can, run out of the location; if you can't, run, then hide and barricade yourself in; and as a final resort, if cornered or found by a shooter, fight.

The questions went on for a long time, with people asking how they could protect their children, for example – all the concerns that flood people in a state of panic. I remember later on, walking near the Place de la Concorde, where the Christmas market is normally thronged with people. All I could see that night were hundreds of armed French police and army, patrolling the deserted streets.

———◆———

In the late summer of 2016 my old sergeant major Davy asked me if I wanted to come work for him and another ex-Ranger, Vincent, who ran a security company in Afghanistan, Crean International. Private security companies in these conflict areas were normally employed by various businesses or non-government organisations to provide intelligence, training, risk consultancy or direct security for the company locations and employees. Some security companies are huge and operate almost like a small private army for its clients. I was employed as a security manager and would manage our teams on the ground consisting of security guards, protection officers, technical teams, searchers and the hundreds of other personnel on our sites keeping everyone protected. We would also carry out threat and risk assessments on potential attacks and possible weak points in a company's daily operations along with providing guards, intelligence updates and threat warnings.

Crean would be providing security for Roshan, the Afghan telecoms company. Roshan employed over 900 Afghans, 20 per cent of whom were women in a male-dominated society where unemployment was at 40 per cent and work was hard to come by. The company is a vital communications link for millions of subscribers in all 34 of Afghanistan's provinces, adding over 30,000 jobs to the economy with the services it provided. Roshan operated many charity and redevelopment projects; it was also a leading social enterprise, working to rebuild communities decimated by years of war. The company was part of the network of the Aga Khan, a prince and the 49th religious and spiritual leader of the Ismaili Muslim faith, who holds the status of a head of state. We would work in hand with other parts of his umbrella organisations, the Aga Khan Development Network (AKDN) and the Aga Khan Foundation (AKF), to create social and community structure for the Afghan people. The Aga Khan also has links with Ireland and owns a number of stud farms. He is as an avid breeder and horseman himself, which is a large part of Afghan culture.

I discussed the job with Sinéad and we agreed that I would go for a month to test out the landscape, as it were. And so I found myself, while on holiday in Valencia, speaking to my potential new bosses on the phone in the bathroom of our hotel suite while Sinéad and the children sat in the bedroom. It all happened very quickly and I flew to Dubai the day after I arrived home from Spain to meet up with Davy in late August 2016.

◆

With Davy holding my hand I arrived in Hamid Karzai International Airport in Kabul where we were greeted by a local company driver who brought us through the multitude of military checkpoints to our office location in the Wazir Akbar Khan area of Kabul. The Crean International office – known as 'the White House' – was a small annexe next to the offices of our main client, Roshan, one of the largest telecommunications companies in the region. The office was in the 'Green Zone' of Kabul, a diplomatic area, with our compound near the German and British embassies.

Along with other expat employees we were housed in a different area to the east of the city in a fortified compound on the Jalalabad road called Roshan Village, which was part of the larger Green Village complex. The compound was guarded by a small army of ex-Nepalese Gurkhas, who would snap to attention when we passed, smile and greet us with a 'Namaste'. I quickly forged a good relationship with these lads. If the shit hit the fan in an attack,

Gurkhas are the guys I would be depending on. These Nepalese soldiers were renowned as tough fighters and for their military prowess, being recruited by the British and Indian armies in large numbers. I had worked with them in Lebanon years before and had a Gurkha on my team on my recce course in Britain, Kam, who I was pals with. A small, strong and humble man, he could also easily straighten up from the ground wearing a heavy backpack while the rest of us had to ask for help.

The Jalalabad road is one of the most bombed roads in the country with constant attacks on anything that looks like a target, whether it is military, government or foreign civilian vehicles. Our 6 km daily commute could take up to an hour, depending on traffic. We called it 'Operation Certain Death' because we had to do it a few times a day and changed the route as much as we could in order to avoid detection.

◆

Crean International employed hundreds of Afghan watchkeepers and staff who were located on telecommunication sites all over the country. My job was to manage operations of these men and women as well as providing protection for our foreign staff and intelligence for other clients. It all seemed incredibly complicated from the start with a lot of players and moving parts, along with the constant bombing and attacks in the city, much of which is not reported outside the country, mainly because the frequency of attacks meant that only the very large body counts grabbed international headlines.

I was taught the ropes by my bosses as well as by two ex-British military men, Dave R and Roger, who had worked in Afghanistan for years. These guys were outstanding individuals with a huge knowledge of the area and the routines, which they showed me over the following weeks. They also fully accepted me into the team and almost treated me like a brother, keeping an eye on me, having good craic and winding me up in the process. I loved working with them. Dave R had been employed by a brewery for a while and used to brew his own beer and wine in a bin in his room. On down time it had the required effect and even tasted pretty good. We had our own labels and even made some Guinness-like stout at one stage, using an intravenous set (IV) to bottle it.

We observed the Islamic week, with the weekend being Friday and Saturday, and Sunday the first day of the working week. Fridays and Saturdays were spent in our residential compound managing the teams nationwide and staying in touch with operations. We spent the time catching up on admin, checking emergency kit and medical bags, cleaning weapons and training. On Saturday mornings we would gather for breakfast and cook up what we called a 'full Anglo-Irish breakfast', usually headed up by Dave R. I would bring back Clonakilty black pudding, bacon and Sinéad's home-baked, frozen brown bread. Dave would have Cumberland sausages, bacon and HP brown sauce, and we would source the other things locally. It was a catch-up with six or eight of us, mainly security team guys of various nationalities sitting around a table in our bosses' room, which had a kitchen and was a magical piece of home for a few hours a week. Our company dog Georgia would hover about looking for the leftovers, having learnt to open doors and gain entry as required.

Afghanistan is a very complicated place with several languages and various tribal or ethnic groupings that didn't get along, and that was just the government. A country renowned for being tough to crack, never having been fully controlled by the many who tried. The Persians, Alexander the Great, the Mongols under Genghis Khan, the Russians and the British have all tried and failed, as NATO had learnt most recently. There were now multiple insurgent groups (and not just the Taliban as I had thought before I got there) including ISIS, HN (Haqqani Network) and Al-Qaeda. The main threat besides bombing was kidnap of our clients for ransom, which was big business here. Then we had the regular bandits and criminals and finally the Afghan police and army that were also factionalised, and often not above criminality either.

There was no typical day but of course there was a routine, which revolved around arranging transport, security teams and routes not just for us but for our international and Afghan staff. Everyone is a target in Afghanistan so our international staff can't just walk out onto the street. Anytime they go anywhere their route has to be prepared and escorted by a security team. Threat analysis and intel would dictate routes and whether a journey could be made or not, with constant reports on potential attacks on this street or in that ministry or location – it never stopped. If the risk was too high we would shut up shop and nobody travelled anywhere, because you don't take risks in Afghanistan. I'm sure that some of the staff saw us as a hindrance to their business and they would sometimes get the hump or complain if we refused a journey. Nobody wants to hear from security – well, not until the shooting starts and then suddenly we are in vogue.

During the day we would manage our operations countrywide from our main office but we also had two 24-hour operations rooms running. Like any company we had all the normal daily work with personnel and routine and where possible we would get out to some of the locations we protected to support the teams on the ground. We would return to our home base in the evening along with the internationals and continue to monitor all operations as required remotely. Where we lived was like a motel and we had a gym and other real-life comforts like a coffee shop, bar, restaurants and shops all within the protection of the perimeter like a proper village.

The Afghan head man who managed our day-to-day activities was called Mr Nawab. He had a mysterious past and had spent some years in the former USSR, I'm sure in an intelligence post. The lads joked that he was KGB. He was an honourable and brilliant man with a wealth of knowledge on the country and his finger on the pulse for information. He had spent his adult life protecting people in a country that has been in a state of conflict since the 1970s. I completely respected his direction and if he said to do something, that's what you did. I enjoyed working with him and hearing his stories of life in Afghanistan and its history.

Some nights you would hear a blast or get a message from our NATO intelligence sources of an incident. Then, usually within minutes, my phone would ring and Mr Nawab would begin with a 'Good evening, sir' – even at four in the morning. He would tell us what it was, who it was and the impact on our people or locations and would normally also add what the problems were and how he had already fixed them. He would always finish the conversation with a 'Thank you, sir' after which we would usually send out an alert and account for all our people. We would then have to report any situations to our employers and bring in any other groups we might need to assist. The incidents varied: for example one of our watchmen was shot and injured by a US patrol one night on a hill in Kandahar. The US team moving near a comms mast we were protecting made some noise and wounded our guard when he lit them up with his torch to investigate.

I started to learn Dari, the most commonly spoken language, which Mr Nawab and some of the other Afghan lads were happy to teach me as I went along. I was also interested in Afghan culture and history, which immediately helped me forge bonds with our local team. Afghans are proud and warrior-like people who stand on honour and family. I found that a clear and authoritative approach worked best. I had to be direct, sometimes even harsh, which is the way it is there.

◆

At the end of my first month in Afghanistan I was looking forward to going home for a break. I was in the Emirates queue and about to board the flight out of Kabul to Dubai when my phone rang. It was Sinéad. She was in the early stages of pregnancy at the time, and the baby's heartbeat had stopped. It had happened to us once again, losing yet another life. We were devastated. I spent my first month at home trying to pick up the pieces and being there for Sinéad and our family.

Sinéad and I talked at length over that difficult time, and decided I would carry on with the job in Kabul, slotting into a routine of doing two months in the country and one month off, which I did from that autumn onwards. It doesn't suit some families but to us this was our normal: we were accustomed to me on the road and it worked for now. I was, however, very aware that Sinéad had the worst of it because every time I went back to Kabul there was a chance I wasn't coming back, so she worried of course. She didn't watch the news on Afghanistan, and I didn't tell her much about what I was doing on a daily basis. I just kept it light, with anecdotes and stories of a normal nature. If I had been involved in something, I would call her as soon as possible to let her know I was OK. She was amazing, to be honest, and made it easy for me with her endless support, constant organising and positive attitude.

———◆———

One day that late autumn, shortly before I went back to Kabul, I met up with my old friends Killer and Baz and some other pals for lunch. I'd arranged to pick Killer up from his apartment and when he first opened the door to me, I was shocked at how much he was fading. He carried an oxygen bottle with him and looked like an old man as he struggled through lunch, still projecting positivity, of course. After our meal I brought him home and accompanied him into his building and up the four flights of stairs, which he was adamant he could manage. I watched him struggle up each step like a student on a mountain march on selection. He didn't want help, and it only really hit me then that he wasn't going to survive this illness. I had been very much detached from it, by being away, with Baz and others carrying the can daily in support of Killer. I went back to Kabul again and thought about him on the long journey and of course of my own family and our particular loss. I am really good at putting things in my back pocket so that's what I did when I stepped off the plane in Kabul.

I would be spending that Christmas period in Kabul, during which I was in constant contact with Killer leading up to it. One night he sent me a text that he was fighting hard, and I replied that I was looking forward to seeing him when I was next home. He replied he hoped to see me too, but he never did. Killer died shortly before Christmas. I was hit hard, to be honest, especially as I wouldn't be there to pay my respects to him with our comrades and friends.

———◆———

The Afghan winter is harsh and cold, but on clear, bright days you can see the snow-capped Hindu Kush mountains that almost surround the bowl in which Kabul sits. The air quality is terrible, and most nights the sky is so full of smog that you can taste it in the back of your mouth as people burn everything to stay warm, wood and coal being an expensive commodity. Children push wheelbarrows along streets full of litter and plastic en route home for the fire.

I was in country for Christmas with another colleague, Neil, covering the holidays, which was OK as I was due out for the New Year. My boss Davy flew in just before Christmas for 24 hours from Dubai where our head office was. He didn't come to work, he brought the complete makings of a Christmas dinner and cooked it in his kitchen for us, so Neil and I had a proper homecooked Christmas feast with all the trimmings. Not too many bosses would do that for you and show that level of empathy, kindness and support.

The buzz of helicopters is a constant in Kabul. They pass overhead day and night, mainly US forces who rarely travel by road. I learnt very quickly that here, life is expendable. Attacks happen daily, in the city and elsewhere, with the civilian Afghan people taking the brunt of the damage, with the death rate in the thousands that year, 2016. The fighting between Afghan government forces and Taliban or ISIS was ongoing in all 34 provinces in the vast country of 51 million people. Some areas were completely controlled by Taliban and the attacks in Kabul were seen as a show to the government that it was not in control. In the earlier days bars and restaurants were frequented by internationals as attacks were rare in the capital, but that all changed after some major attacks on hotels and social gatherings.

The insurgents wanted all the foreign troops out of the country and the restoration of a fundamental Islamic state along Taliban teaching lines. One of my Afghan workmates told me that when he was a boy in Kabul in the nineties, at half time during a soccer match the Taliban would carry out executions in the centre of the pitch, after which the match would resume. The main risk in Kabul were magnetic mines or bombs, which were stuck onto the side of a target vehicle, regardless of who was in it, be it a school bus or a family vehicle. Terrorists on motorbikes scouted the city for opportune targets and would indiscriminately hit police, military or government vehicles and foreign nationals.

As Western security contractors it was important that we keep a low profile and that's why we chose a more undercover approach to getting around the city. Some teams, mainly US, had armoured jeeps covered in comms aerials and stood out for miles in the city, as Americans tend to do. Armour is great for bullets but the tactic of terrorists was huge, explosive attacks. We used nondescript local vehicles with our own Afghan drivers to go everywhere and blended in with the mainstream. Road travel outside the city was a non-runner. The Taliban would stop buses in the regions and shoot Afghan men, or hang college students, never mind Westerners. Several Red Cross staff were beheaded when captured in a remote region in the north while I was there.

Our main job in Kabul was to anticipate attacks and to avoid contact and danger at all costs, for both ourselves and the people we were protecting. The intelligence on a daily basis on suspected attacks or suicide vehicles was immense, with a constant update on new threats. We had a few close calls, of course, but we were getting it right and keeping people safe, not just our foreign nationals but the thousands of local staff who worked for both Roshan and some of our other clients, including banks and non-government organisations.

Large-scale complex attacks were now the regular tactics of the insurgents, with a large vehicle bomb being the entry method, followed by several suicide attackers with explosive vests following on foot. I had to lock everyone down in the main Roshan office one morning as the military hospital nearby was attacked by terrorists posing as medical staff. They killed 51 people, some of whom lay helpless in their beds as their throats were cut. The battle to regain the hospital went on all day with my team and I prepared for us to be the next target, such was the situation.

But one of my most intense experiences in Kabul took place end of May 2017 during Ramadan, the holy month of Islam – unfortunately not observed by the insurgents, with power and hatred all they care for. On my commute to the office with Davy, a few minutes from our destination, a blast went off that I initially thought was right next to us, it was that loud. In fact it was a huge truck bomb, the largest ever in Afghanistan, close to seven tonnes of explosive in a tanker truck meant for the German Embassy just across the street from Roshan and our offices, which were not the target but collateral damage. I was in immediate contact with M, the Roshan security director and former ARW colleague on site.

We arrived at the blast site and there are no words to describe the devastation and aftermath. It looked like a post-apocalyptic movie scene. For a three-kilometre radius the streets were covered in the glass of the shop fronts and windows of the buildings that were smashed in the shockwave. The huge blast walls that surrounded our office were gone, as a blast that big will create an overpressure that will go over if not through structures.

The carnage that greeted us was indescribable. More than 140 people were killed in the area with hundreds if not thousands injured. A rescue operation was mounted to try and find survivors in the collapsed buildings, which quickly turned into an operation to recover bodies when no one was found alive.

Roshan lost more than 30 people and over 80 were wounded, many seriously. Among the dead were two of our Crean guards, killed at their posts, and two of our Afghan protection officers, one of whom was a friend, Rashad. He often gave me gifts to take home to my family and he loved Irish tea, which I would always bring back for him.

Mr Nawab and several other staff were also wounded but he remained and was a major part of the clean-up and recovery operation. The courage of our Afghan team was considerable. I was humbled by how much they will work and continue to protect a location that was in essence destroyed.

The beautiful garden in the centre of the complex, where people could sit and enjoy some peace and quiet, was now a field of rubble from the collapsed buildings. We had to bounce back quickly and within hours of the blast held an executive crisis meeting to determine a course of action. We needed to firstly secure the complex and its assets, to recover the dead and set a timeline to be back operational as a business – which is tough, I know, but that is how it has to be. We set up work teams to search for people and to secure business critical assets, which was very hard in the June heat with Muslim workmen fasting from dawn to dusk in Ramadan, not even able to take water.

To add fuel to the fire there was a huge demonstration outside our offices the next day from civilians in protest for the failure of security forces to keep a so-called safe zone protected. The blast went off at a pick-up point for buses and cars, mainly taking children to school. Several of those children were killed instantly. We now had a 20-foot high canvas screen, where once our blast walls had stood, so we could see the silhouettes of the vast crowd gathering outside our thin perimeter.

The Afghan security forces used tear gas, which gassed us in our compound also, and finally broke up the gathering by firing PKM machine guns in the air from vehicles as they pushed the crowds back, killing another 12 people and wounding many more. We took some fire also; the work team and staff I had with me I pushed into what was left of our basement for protection. The city descended into even more chaos for the next week with the death toll up, more bombings of locations and mass protests and civil unrest – so much so that we began getting our country evacuation protocols ready to get our foreign nationals out of the country that looked like it was going completely under.

The CEO of Roshan, Karim Kosha (known as KK), flew in and as part of his support to see his beloved company and staff, he visited every one of the families of those killed in the blast, openly weeping as he did so. We put a protection team on him and I travelled with him in his B6 (a 4x4 Toyota bulletproof vehicle) around the back streets and alleys of a dangerous and tense Kabul to visit these families.

We would arrive at each home and remove our shoes at the door, as is the custom, and sit on the floor of the main room. We prayed with the families for their loss of not just a family member but also probably their only source of income. I must have travelled to 15 of these homes (Davy escorted KK to the other houses) watching these people pray in their sitting rooms, just like I have seen people do in Ireland at a wake – different religion, same process. It was one of the most humbling experiences of my life to experience the emotion of families in this raw state.

I felt so helpless that I hadn't been on site at the moment of the explosion, but I realised in time that even if I had been, I would not have been able to stop it. It's important to be able to rationalise events like this and reconstruct them in your head in a logical manner to learn lessons. If you can't do that you risk messing up your head, which makes you ineffective and a liability to yourself and your team. We buried our dead, reset what we needed to do and got on with it with purpose and commitment. To be honest, part of me wanted to pack up and just go home, I just had to ride that out and bring it down to basic routine, completing small tasks to build confidence in myself as well as my team.

The management team of the company, along with our support, was able to rebuild the company quickly so that we were almost as functional as before the blast. It wasn't easy but with the right can-do mind-set we were able to redirect our team and restructure our work. We set up a new office in Green Village, so my death-defying daily commute on the most bombed road in Afghanistan was reduced to a five-minute walk, where I didn't even have to leave the compound.

Shortly after the bomb I went on home for the month in my rotation and a much-needed break. It had been a pretty intense time, and I needed to make sense of it all in my head. I spent most of the month with Sinéad and the children in my brother Ger's summer house in Spain. We had a great holiday, where I mostly played on the beach with the children, doing regular family stuff. It took a little time to untie the knots in relation to the blast but I was eventually able to understand and make sense of it. I was in good order when I left my family there to travel back to Kabul.

———◆———

I returned to Kabul in the summer of 2017 and from there I was quickly redirected to Pakistan, where we had a sister operation that needed support. I would leave for Islamabad in early August and arrive in the unbearable humidity of the wet season in Punjab state. This time the operation would be business- rather than security-oriented. I would work with Gus, another former Irish soldier, and a team of Pakistani technical guys on the project, which would see me travel all over the country setting up electrical-engineering workshops where we regenerated batteries in Rawalpindi, Karachi and Lahore.

I lived in a suburb of Islamabad called Bahria Town, where the locals were both very respectful and suspicious of me at the same time. It was a refreshing change from the tight security of Kabul as I had a certain freedom of movement, although terrorism was still a real threat, along with being monitored as a foreigner by Pakistani intelligence.

The population of Islamabad is 14 million, so the commute to the city for meetings and work was something to behold – a sea of motorbikes and tuk-tuks (covered tricycles). I went to many meetings with my good friend Aamir, CEO of our company, and sat through hours of bargaining in Urdu.

The meeting process in Pakistan is long and formal and broken into distinct phases, denoted by the serving of sweet, milky tea, as is the tradition. The first Urdu words I learnt were 'No sugar, thank you.' The first cup of tea is a welcome to the meeting; the second cup is to confirm the business deal; and the third and final cup is to welcome you into the family. Every meeting was like this, going on for hours. Only once in the year I worked in Pakistan did we not follow meeting protocol, when a 6.1 category earthquake erupted during the meeting. The five-storey building we were in wobbled like a Jenga Tower, the floor buckling under our feet. When the aftershocks subsided, we got out from under the tables and just carried on as if nothing happened.

I had a team of around 20 people to manage in various locations. Pakistani society is very much based on the Islamic honour system along with a rigid caste system – not without its challenges in a vast country of huge cultural diversity. The people I met were gracious and curious. I was there for the Muslim feast of Eid ul Adha, welcomed into the families of some of my workmates like I was one of them, and shown incredible hospitality in a society where pious humility is respected and practised above all.

I would be going home for Christmas and was enjoying the thought of spending it at home to see the children on the day. I appreciated the differences in Pakistan and the fact that I had a certain freedom of movement I didn't have in Kabul. The work was also completely different with security not my main effort, which saw me experiencing new things and different skills. I would stop in Kabul first for a day or two to catch up with the lads and go to our Christmas party in Kabul.

CHAPTER 10

REORG

spent Christmas 2017 in Ireland with my family. Having been away for so long, I was determined to enjoy my six weeks of leave, but it ended up being tough on us all. After a few weeks of exhilaration at being home I began to struggle, and my routine and discipline fell by the wayside. I was drinking most days and sleeping too much. Very quickly, I fell into a rut. I was in a bad place mentally and struggling to maintain a purpose. Sinéad noticed, of course, and I had to give myself a kick in the ass to get my shit together, which I did by filling my days with more structure and routine, and this eventually got me back on track. I learnt that the following works for me:

- **Physical activity.** Whatever it is, at least five times a week for 30–60 minutes.

- **A good diet.** Cut down on processed food and remember, sugar is the enemy. Drink two to four litres of water a day and eat lots of fruit and vegetables. Don't skip breakfast and hide the biscuits until you've earned one.

- **Sleep routine.** Try to get seven to eight hours a night and go to bed early enough to unwind for a bit and de-fuzz your brain. Don't look at your devices. Unless you're on emergency call, leave your phone in another room. Get up early enough to have time to start your day well and not be in a rush. If you start on the back foot it's hard to get back on track.

- ▶ **Purpose**. You need a purpose to keep you engaged and motivated, whether it's work or anything else. Have a plan and challenge yourself.

- ▶ **Fun.** Craic is essential, so find something you enjoy and have a laugh. Humour is so important and takes the edge off stressful situations.

- ▶ **Mind-set.** On the tough days, stick to the good stuff. Pick three positive aspects in your life and concentrate on them to keep you going.

- ▶ **New stuff.** Be open to trying something different and engage your brain to learn something new. Variety gives you a buzz and keeps things interesting.

You need to plan your day or week so you remain in good order and don't dwell on things too much, and a balance of physical activity, mental stimulation and positive relaxation are optimal. The army term 'reorg' is a useful one here: it's short for reorganisation, an activity carried out to reset and get ready to move on, and we all need to do it.

I returned to Pakistan in early 2018 and spent a large part of the following year there, interspersed with visits home to Ireland. I was happy enough to be finished with Pakistan in the summer of that year, as we were winding down a lot of the operation, and I returned home to Ireland, where I would remain. I had the whole summer with my family and took a complete break from work to really get back into the loop of family life. I was always cognisant that Sinéad and the children had their own system and it was up to me to fit into it and not try to take over.

———◆———

In October of 2018 I got a call from an old colleague, John, to ask if I was interested in making a TV show for RTÉ, where we would put civilian recruits through a simulated 'selection' process over the period of a week. I initially said no – well, I think I told John to fuck off, only half joking as I thought about the fallout and the fact that I would be sticking my face on the telly for all to see. I had spent my entire adult life working in operational security and restricting the information I shared and now we four were going to become the face of our former unit for all to see. It was a huge responsibility to take on and I was completely aware that it could go all horribly wrong for us. But I mulled it over for a few weeks agreed to a number of meetings with the production team, in which Jamie Dalton, the executive producer, and Jane Taffe, the producer, and I talked about the project and how feasible it was for me. Along with the other three DS, Ger, Staff and Alan, we were adamant that we would follow the principles and format of the ARW course as closely as we could to make it as real as possible while not giving away any secrets. The three men I would work with were the main reason I said yes. I had put my life in their hands before and each of these guys I would go anywhere with. It's not what you are doing, it's who you are doing it with, and I couldn't have had a better team.

In the end we decided to go for it and so, in the winter of 2018, we did a test run in Wicklow. Jamie hired six unfortunates from a gym who we filmed as would-be students, being put through a physical event by the four of us in the freezing rain on the side of a mountain as we crawled them along a small stream. The production team thought it was great and RTÉ went for it also. We planned to start filming it in early March the following year. I just had to complete one final rotation in Kabul over Christmas, which would see me flying out in mid-December until the end of January 2019.

◆

Christmas in Roshan Village was relatively quiet, mainly because nearly all of the foreign staff had gone home for Christmas. I was the only senior security manager in country and would cover Crean operations as well as acting as director of security for Roshan, with the support of my Afghan teams of course. Besides our company dog Georgia, a beautiful Romanian shepherd who slept in my room, the only other expat with me was Michael from Brooklyn, who was in his sixties and had spent years in Afghanistan. He was an events manager and a great craic with his stories of his experiences as a gay BBQ chef in Texas and a TV chef in Poland.

As snow fell that Christmas Eve I had to lock down the compound. We had had reports of an attack that was to be mounted on an Afghan government location nearby. We were on the highest alert, with no movement in and out of the compound, people put into secure locations with the security team and the Gurkha army fully deployed and prepared for an imminent attack.

Intelligence reports had indicated that terrorists might carry out an attack on a foreign compound, which they often did to mark the holiday, so we couldn't take any chances. The attack down the road was a complex suicide assault on what was the nearby Afghan Ministry of Martyrs – the irony was not lost on me as the time ticked towards midnight and a white Christmas. In my head I could hear Bing Crosby singing; in reality I could hear the explosions and gunfire of the battle raging in the distance. I wanted them to just piss off so I could have a glass of homebrew and see what Santa had brought me – I had gifts from the children, which I promised not to open until Christmas Day. Soon enough, however, the security forces managed to clear the ministry, but only after the terrorists killed dozens of people. It was a bittersweet Christmas Day, opening my presents in the knowledge that lives had been lost.

———◆———

The rest of Christmas passed by with little activity, me training in the gym or walking Georgia, who believed she owned the compound and wanted to play with everyone. New Year arrived safely enough, and I passed it with a couple of Scottish colleagues from another company in their kilts for the Hogmanay celebrations. It was starting to get busier now as a lot of the foreign staff were coming back. Business was on the up for Roshan. We had four main sites in the compound, which we shared with several other organisations.

These included the main offices, a call centre, a field office and staff accommodation with around 60 rooms.

Vincent and Davy were also back after the break and living on the corridor where I was housed, which was also our safe room, with its reinforced walls and blast doors. Davy had just moved from his previous VIP room where we had cooked weekend breakfast and which featured a dangerously exposed upstairs bedroom. I would train each day in our gym across the corridor or enjoy a game of squash with Vincent, after which we would have dinner together.

◆

The blast that threw me into my room, which I mention in the prologue, was the result of a suicide truck bomb detonated just outside the rear perimeter of our block of accommodation. It was 7 p.m. on 14 January 2019. The two tonnes of homemade explosive (HME) had a yield of the equivalent of 1 tonne of TNT and went off 75 metres from where I stood in the corridor. Blast waves follow the path of least resistance and this one came along the corridor like a freight train.

I grabbed my body armour and made sure I had enough magazines so I could do a bit of damage if I had to, along with landmarking or feeling for my medical kit, which I knew I would surely use. Like in SF, all the kit I need is attached to the body armour and prepared, so once it goes on I have all I need to fight. I quickly sent a group SMS message to all staff to stay in their rooms and lock themselves in their bathrooms. Since these were entirely covered with ceramic tiles, it was the safest place for now. I also didn't want them moving about the compound if we had terrorists running about.

It was pitch black and eerily silent as I composed myself and took a few tactical breaths before I went out the door into the corridor, now filled with dust and debris. I scanned towards the main exit and stepped tactically forward, my AK at the shoulder, 'eyes up', scanning the dark. I was listening for the sound of gunfire, which I couldn't really hear because my head was ringing from the blast. I was going to go to our main vehicle gate for Roshan Village, which I believed was the attack point for a follow-up and the place I had to secure first.

I had to cross an open courtyard quickly as I expected fire at any moment and moved towards the Gurkha guard post. As my eyes focused to the darkness, I hoped the gunfire would be from the front if it came, so the ballistic plates in my armour would take it – the soft side armour wouldn't stop the 7.62mm bullet used by the terrorists. I was also without my helmet – I couldn't locate it after the blast – so my head felt extremely exposed and even fatter than usual.

The layout of the compound looked different as I passed the damaged buildings pushing towards the main gate area, scanning for targets as I went. I passed through what had been the guard room and gathered up two or three guards who were hunkered down there, pulling them with me to clear the main gate, which we did. The double-gated air-lock steel gates were just about intact, if barely hanging on, so along with the guards we secured the gate and I positioned them to cover any movement through them.

The call centre was immediately inside this area, along with our vehicle park, which now resembled a scrap yard with the damage to the cars there. The call centre had around 75 Afghan staff, so along with our duty operations guy Nawid we cleared the two-storey building and pushed the staff out of there and back towards the safe corridor. People jumped out from every corner and from behind desks when they heard my voice in the dark, some of them coming close to being shot, to be honest. It was chaos, but my training stood to me. In this scenario people go to pieces and grab at you like a person drowning. I had to enlist some staff to physically drag others from closets and assist the injured to bring them with us back to the safe area.

I moved back to the corridor where my room was, with Nawid's help, pushing the 75 staff into the location and started to take a count of who I had. Davy and Vincent were there also, having been delayed as Davy had been knocked unconscious in the blast. We now had to clear the accommodation block and locate our 36 foreign staff in their rooms in a network of long corridors. Vincent took some guards, as did I, and we wanted to ensure that no terrorists were in the block.

Just before I stepped off, I thought 'Georgia!' I ran to her kennel just outside our block. The gate was blown off and with the light on my weapon I quickly scanned the little yard, then spotted her cowering under a sheet of plywood. I quickly gathered her up and ran into the safe corridor, where I put her into a room that still had a functioning door. I calmed her down for a few seconds, cut the top off a plastic water bottle for her and quickly checked her for wounds, but she was unharmed, thank God, although terrified.

Although the Gurkhas were excellent soldiers they lacked initiative, so the four I had just stuck with me as we methodically cleared the rooms and brought the staff we found back to the safe corridor as we got them from their rooms. The safe room was protected by guards under the control of a South African colleague, Craig. The Gurkhas were brave and it was a great comfort to have more guns on my shoulder as we cleared room after room, not knowing what waited for us as we stepped in each door. The rubble-strewn floor was hard to negotiate and we had to constantly climb over collapsed walls and furniture. It was hard to see anything in the darkness and the dust that hung in the air like fog, which the torch light didn't penetrate, so I was going on instinct in some rooms.

When we got to the far side of the accommodation block, I could now see where the blast had been and the most damage was to rooms in that location. It was flattened. The three layers of external blast wall were blown onto the rooms and the wall of the building was gone. It left a large gap in the perimeter on which I posted some guards, to plug the threat. I met a Kenyan staff member who walked out, covered in dust. I was amazed and delighted he was safe.

After we cleared the block and I moved back to the safe area, we immediately took control of everyone in the safe corridor. I shouted loudly, asking everyone to be quiet. I could hear the sobs and prayers and the crying of the injured in the hallway, illuminated by the occasional torch or phone screen. 'You are safe now. I need you to follow our direction and do as we say,' I explained to my captive audience of over 100 civilians. For a second I suddenly felt the reality of the responsibility I had for these people's lives as the commander on the ground. I'm not a robot and was emotionally involved here, knowing a lot of these people well having lived with them for a couple of years, but emotion will get you killed so I focused on my process of control and composure. We started to account for people in our protocols and soon discovered we were missing at least ten.

Some people were in the larger adjacent Green Village compound but we were still missing three people in particular, two US nationals and an Indian lady. This was a multitask situation at its best and I had to trust others to complete tasks I hadn't the time to do. I delegated tasks to some of the managers I had in the corridor as I needed to find the missing staff in the block. We got a group to care for the injured with the medical kits we had in place – it was mainly light injuries. I also got a senior director to start going through the call list and find exactly where everyone was and to report back. I did the same for the Afghan staff and Davy took control of locating the missing in the block as I went to liaise with the support elements inside and outside the location. It was a gift to have Davy and Vincent on scene – they are my two bosses and had so much more experience, but understood I was in command on the ground and quickly stepped into a support role, developing a course of action and helping out any way they could. It would have been a different outcome, I'm sure, had they not been with me.

The danger now was that there were lots of armed groups moving around the compound, various international organisations and Afghan security forces, so blue on blue (friendly fire, or shooting each other by accident) was a major risk. My phones were now also hopping from people both inside and outside Afghanistan looking for updates. I hadn't time to deal with them, so ignored the ones who couldn't do anything for us. Senior executives and anyone with a title were wanting reports from me, which I completely disregarded. In a crisis you have to prioritise what is urgent only: everything else will be done later so block out what is not immediate. I created a link with our support network in country and Pete Tait (another Irishman) in particular, security director of our sister organisation, the AKDN. He immediately started to set the conditions to get us all evacuated to a secure location. He also understood that I was in a pressurised environment and didn't call me, but waited until I had a minute to call him, when we would update on the situation for evacuation and a timeline for what we had to complete. He was a lifeline.

Because of the surprise nature of the attack, we were on the back foot and everything we did was reactive. Meanwhile, the terrorists had seized the initiative. Our first priority was to secure the compound, which we did, dispersing the guards and using the quick reaction force from a support group called Shield within the complex. Next was locating all of our people. Vincent and his team came across discarded weapons and equipment at the end of one corridor, left by terrorists, so they were in here and could attack at any second. We'd been lucky up to now: the weapons had grenade launchers and the vests contained grenades; we didn't understand at that moment why they weren't used on us.

The three of us were searching the shattered buildings when Davy called me as he had located two of our missing staff in a corridor close to ground zero of the blast. When I got there I could see under torchlight that someone was trapped in the ceiling above our heads and another figure lay on the ground, covered in rubble. Davy uncovered the face of our missing Indian colleague, Shipra. I stepped in and did a quick primary survey on her to assess her condition. I had never met her – she had only arrived back from leave late the night before. We cleared the rest of the rubble from her and I secured her airway. Although covered in dust and dirt she didn't have a mark on her or any obvious injuries, so I stabilised her to move. We needed to get her to medical support in Green Village immediately as we feared serious internal injuries and we expected a follow-up attack at any minute.

We got her on a stretcher and Vincent and his team quickly got her to the casualty collection point in Green Village which was the squash court, where medics worked to save her. Unfortunately she died from her injuries, so we didn't save her as I had hoped. She was a beautiful young Indian lady who had come to Afghanistan to work for the Afghan Institute of Civil Society (AICS). Instead, she died in the cold with people she didn't even know. We put the thought of her in our back pockets for now and moved onto the objective we had to achieve.

The person trapped in the ceiling was Michael, who was giving out and bitching to us, as was his New York manner. He was a great old friend of Davy and we both hacked and pulled at the masonry in the dark to get him out of what had been a mezzanine floor above us. He was fully clothed but had no shoes on, so when we eventually got him to floor level, we realised he needed footwear in the rubble. Davy fished about the floor and pulled a pair of trainers out of somewhere and, passing them to Michael, told him to put them on. Michael, a stylish gay man, said in his best Brooklyn accent, 'I'm not wearing those shoes, they're awful!' Davy and I both said at the same time, 'Put on the fucking shoes!' We needed to get him to the safe corridor quickly.

We were able to account for all our people except for one, Mano, the CEO of First MicroFinance Bank, so we moved everybody to a safer location in Green Village, where we had more protection and facilities. Along with Pete Tait, who was off site, we planned an evacuation to the secure Serena Hotel (a five-star hotel in central Kabul), if we could just get the relief convoy here, a task in itself.

I could see people were starting to break down at this stage, hours into the attack. Some had no coats on in the −7° cold and although we were technically inside, there wasn't a window left in the complex. We arranged for warm clothes and got people to prepare an escape bag, which some already had, as we had trained for this and carried out practices for just such an event. One man in particular was very bewildered as he had been in country for just a week on his first term working for Roshan. I had conducted some hostile environment awareness training (HEAT) with him just days earlier, so he had his escape kit packed. He commented that he didn't think he would be using the training this quickly.

We continued to update the staff, making sure they had water and a little food, and handling the situation as best as we could. We were very deliberate in what we were saying and doing to give an impression of control. There was the occasional bit of panic as people overheard information from other sources or from their phones. We continued to search for Mano, taking turns digging with our bare hands in the collapsed building, which was useless – it was dark and well below zero, we would need mechanical support to find him. We were also continually liaising with Pete, who had arranged an evacuation convoy for the foreign staff while another colleague, Shuja, organised a convoy of vehicles to take our Afghan staff home.

We lined up everyone with their hands on the shoulder of the person in front and walked them in a long line to the last surviving gate in the complex; all the other entry points had been rendered unserviceable in the blast. The language you use with people in this situation has to be simple and direct with as much empathy as possible. You have to physically move them at times, as they are suffering from shock and the trauma of it. Shuja looked after the local staff and we had them all at home safe by the early hours of the morning.

I called the names of each of the foreign staff. We had to reorganise them one by one as we lined them up and moved them into the ten armoured B6 vehicles we had with Pete's team from AKDN. People were now struggling. The closer it comes to rescue, the harder it gets, with several of them in tears and hugging me as I checked them off to a vehicle. We also brought Shipra with us in an ambulance Pete arranged and gently placed her in it with as much respect and dignity as we could. We did this very discreetly out of sight of the main group, as they didn't need to know and definitely didn't need to see the body bag.

Vincent and Davy stayed in RV along with Craig to keep the search for Mano going as best as they could. We evacuated the complex by 2 a.m. and drove the deserted streets, arriving at the

Serena Hotel, where a medical team was waiting to help the ten injured people and give them rooms to rest in. I even got cleaned up myself by an angry doctor who was adamant that the scratch on my head needed attention, so I just went with it.

I was in constant communications with Roshan senior management, who would later evacuate all the staff involved out of the country. I got a room for myself at the Serena Hotel and just sat on the bed for a minute, thinking about what just happened. I was covered in dirt and I needed a shower badly. I had been convinced at the time of rescue that we had saved Shipra, and I hated leaving Mano behind. But I quickly sorted myself out and arranged for the AKDN security team to secure the survivors so I could return to RV and start working.

Before I left the hotel the next morning, I had breakfast with some of the survivors and a guy asked me straight out, 'How are you calm in a situation like this?' He also went on to tell me that our presence made him feel safe and he thanked me, as did all of the eight people at the table. I left for the blast site. It was another humbling experience, with people I would never see again, and confirmed to me that if you can manage your own reactions and composure first, then you can work on everything else.

Once again we got the wheels in motion quickly, repairing and replacing all the blast walls and gates to protect what we had. In Afghanistan bomb repair is an everyday occurrence, so the clean-up happens quickly. People get on with their lives. Meanwhile, Davy controlled the search for Mano, who was finally found just before 5 p.m. the day after the blast after hours of digging with machines and close to 24 hours after the explosion. The security team recovered his body with respect, before removing him. In all, seven people were killed in the attack and over 80 were injured – not all in the complex but families in their homes nearby also. The compound was attacked again later that year after I left, with 16 killed and over 100 injured.

After a few days of getting to grips with what we had to work with we started to get into our routine as best we could once again. I got back into training and Vincent and I went to play squash to focus on something else for an hour, but when we stepped into the court the first thing we saw was the stretcher that Shipra had been brought in on, which was standing up against the wall where it had been left that night. Another stark reminder.

When people think of Afghanistan it's normally along the negative lines of car bombs, the Taliban and war. But in the midst of this life goes on. People go to work, children go to school, families go to restaurants, play football and carry on with everyday life regardless of the danger. The daily courage shown by these people – the heroic little girl just walking to school despite the dangers of the Taliban – shows the true nature of the Afghan people, who are resilient beyond words. There weren't many days in Kabul when I wouldn't see a child flying a kite somewhere on a rubble-strewn street. On several street corners you will find a wedding hall, a brightly coloured, palatial building standing out amidst the chaos, where families gather for weddings and events, to have fun and celebrate life. This is what I remember about Kabul.

◆

My time in Afghanistan had come to a natural end as my contract with Crean was now finished, although I suppose I had been close to an unnatural end there enough times. I was lucky to have worked for men like Davy and Vincent and the team who guided me was amazing. This attack in RV is something that I will never forget, for many reasons. I mention it because it is the best example of all the things I learnt up to that time, all at one moment. All the things I talk about in this book, that I learnt about myself and others, were condensed into those few hours in Kabul. Of the security team on the ground that night the three most effective people were ex-ARW operators and were the reason that so many were saved – of that I have absolutely no doubt. I said at the start that the greatest achievement of my military life was the protection and saving of lives: that night in the snow in Kabul was it, in essence.

I finished in Kabul at the end of January and returned to Ireland, where I was now determined to reduce my time away from home as much as possible. More importantly my children were getting older and when your seven-year old daughter is writing stories for school about Daddy going to Kabul and missing him, it's time to stop. Two months is a long time in a child's life and every time I came home, two different children met me at the airport who I had to get to know again. Life is too short to miss out on that.

◆

In March 2019 we were getting ready to film the first series of what would be called *Special Forces: Ultimate Hell Week* along with a huge production and technical crew. Staff, Ger, Alan and

I hammered out various ideas with Jamie and his team, but we remained resolute as to what we wanted. The production wanted Hollywood drama – fair enough from a group of film-makers. I remember saying to Jamie that 'Selection is not a spectator sport,' to which he replied, 'Well, TV is!' It was a good answer. We, the DS, on the other hand, although committed to this task, had a higher authority we answered to – the unit – that we had to respect, be true to and not sell down the river.

Initially we were nervous as to how we would be portrayed by production in the final cut of the show and we were forceful in our plan for the programme. I don't think the production team really understood what selection involved until it actually started. Some of them were horrified by our demeanour and attitude to students. For us it was a normal and seamless transition to DS mode, which we had done so many times in the past and really, with no script and no pretence, just us on our size tens, the cameras made no difference.

◆

And so within weeks of a heavy contact in Afghanistan, I found myself standing before a group of civilians, readying them to embrace *Hell Week*.

I looked at them standing in their underwear in the rain. *I'll give you Hell Week*, I thought to myself. To be fair, the production team worked as hard as we did and were consummate professionals to their craft. We saw how the support teams, camera teams and sound guys would slog it out with us over the mountains with as much vigour as I've seen, one lad even in jeans in the mountain snow. The director of photography, Ross O'Callaghan, or 'Rosco', is an expert of his craft with an incredible pedigree. He is an industry legend and a man we were drawn to immediately as a kindred spirit in giving our best at all times. We all worked well together, they got us and what we wanted to achieve from the process and we got them, eventually. It's impossible to recreate selection as the real military pressures can't be reproduced or experienced, but we were happy with what we did as a representation of what it is, with the DS being themselves.

The students who braved the course also need to be mentioned as they didn't really know what they were leaving themselves in for. I had the greatest respect for all of them for standing on the start line that first night, no matter what happened after that, whether they lasted the course or left after five minutes. We still put them through hell because we are fervent that if you want to be associated with Special Forces, even in a TV show, you were going to get it hard and earn that right, which they did, showing people that it's not about being a hard case, being super fit and having disco muscles: it's about mind-set, self-belief and the ability to adapt to adversity.

I remember the production team saying once that after three days we would have so-and-so-many students left and we told them, you can't know that. The beauty is that you just don't know who is going to excel or not. It's all down to them and that's where the main fight is, in their heads. It's hard for a producer to get that and it could all have been over two days in, which confirms that courage comes in many guises and Jamie Dalton is one of the bravest people I know, trusting and depending on us. At times he was fearless, other times we came close to ending his career, but I wouldn't be writing this book only for him.

We completed the show as a team and the editing team under Jamie Fitzpatrick (also our director) did a great job and showed it for what it was, a structured process with no egos, just people who did their best. We had a slot on *The Late Late Show* (my mam would have loved that), which was interesting – the unit reunion was on the same night so we had a bar full of Special Forces critics glued to us. We were happy to stand over it, so much so that when the first episode was aired, we were invited to ARW HQ for beer and pizza to watch it with the unit members. It was a test of accountability and is very important in our culture. You have to answer to someone – if not yourself – for your actions, be they good, bad or indifferent.

◆

Once filming finished in the summer of 2019 I slotted back into the close-protection world, carrying out a few jobs in Ireland, looking after various A-listers visiting the country. Ger and Staff, my old friends from the show, worked for U2 as their security detail and asked if I would be interested in being on the team in the South of France for the summer. I went as part of Staff's residential security team. It was an interesting experience to be involved in something where the threat is completely different and physical security is much less of a problem compared to privacy. The enemy in this operation were not the Taliban or other terrorists, but the paparazzi. I had a lot of experience of discreet surveillance operations in the military so my skills for counter surveillance were pretty good. The paps didn't get a single usable image of our principals or any of their guests in the time we were there.

I spent four months looking after the security and privacy of our principals, Bono and Edge, their families and guests. They are two honourable and good men who have time for everyone, especially their fans, some of whom would sit for hours outside the gates of the compound. They also do a lot of good things for others under the radar and don't bang their drum about it. They had some serious visitors to keep us busy – like the Obama family, who came for a weekend along with a small security team of 30 secret service agents and lots of gendarmerie.

Elton John and his husband David Furnish arrived one evening for dinner – it wasn't exactly low-key, arriving at the gate in a beautiful 1970s Bentley convertible, a tank of a vehicle. Just before it was time for them to leave, his driver tried to start the Bentley, but to no avail. So minutes before the VIPs came out to the courtyard, his security team of ex-Royal Marines and I are pushing a three-tonne Bentley around the courtyard, trying to jump-start it. It was just so mad I couldn't stop laughing, but we didn't want Elton to get a hissy fit – he didn't like to be left still standing …

All in all it was a very interesting summer from May to August, where I learnt a lot about how to protect people who everybody knows and wants a piece of. It was also a great experience to spend the summer in the South of France, where I worked hard with a fantastic team of friends and had one of the best training and health regimes I can remember.

◆

Around the same time I was in France a developer called Greg Kavanagh was starting a new company called Bezzu, which is an app used for buying from smaller independent clothing retailers. It was a great idea, to give these small businesses a platform to compete with the online giants. Greg had seen *Hell Week* and decided that we were the four men he wanted to roll the operation out globally, even though we had no retail or tech experience.

We met him a few times and although I had my reservations, the chance to make a fortune was a big draw, I must admit. The four of us started with the company in September 2019 and began to plan how we would invade the world with this app. I was out of my comfort zone from the start, which is fine, but I had the skills to do it and I was determined to push through with the three lads and the great team we were building globally.

We were tasked with running a fashion photo shoot, the largest ever run in the country, with thousands of items being shot over two weeks. We were given the job some months before and knew nothing about how to do it, so we approached it literally like a military operation. We had to manage everything from the logistics of collections in the shops, shoot location, set up and purchase of equipment, to hiring staff, models, photographers and fire-fight the 500 other problems a day. We even cleaned the portable toilets ourselves. As company directors, people were saying to us 'Why are you doing that?' 'Because it needs to be done!' we'd reply. When you commit to something you have to do everything you can to make it a success, which it was. Our military skills can be adapted to any area – planning, logistics – but the most important is the belief and will to succeed.

We then went to different corners of the globe, hiring good people and building the foundations for the company, in my case in Europe, with a three-year roll out plan. It was a big ask from the start, but we were getting there with some amazing people in the company.

But prompted by the arrival of Covid-19 and the fact that I wasn't really cut out for the job in the first place, I resigned. The experience also confirmed in my mind that my factory setting and gut instinct was accurate, hard-earned as it had been over the years.

Some people are completely caught up in what they have or what material possessions they want. There are things that are much more valuable to me, I rediscovered during that time – life isn't about what you have, it's about what you do. Don't get me wrong, it would be nice to have a house where I had to change gear going up the drive, but I won't compromise my core values. I am much more interested by what people do, what they say, how they relate to others and what they have seen or experienced in life. A saying I heard once has always stuck with me: 'You should judge someone by how they treat people that can do nothing for them.'

CHAPTER 11

AFTER ACTION

wrote the greater part of this book during the Covid-19 lockdown and it has definitely reminded me of some of the skills we need to keep us on the straight and narrow as well as sane. I have probably delved into my tool bag of principles on several occasions over the last year, just like everyone else. It just confirms what I mentioned about being prepared for things to go off-side and have some skills to deal with it.

Today is the last day of January 2021 and I am taken back to the exact time 31 years ago, when I walked towards the main gate in Collins Barracks in Cork and into the rest of my life. Sometimes it seems like a lifetime ago, but it mostly seems like yesterday. I have picked up a few things over the years, usually from better people than me. I was never the smartest, the best shot or the fittest guy in the team, but I was never far off it and I understood how to use that. I am happy to try out things and attempt them with conviction, but I know what I am suited to and what I'm not suited to.

I have since gone back to what I know, security consultancy and close-protection work. I started running leadership training, team-building, motivational and resilience training for various organisations. People were approaching me after *Hell Week* and asking if I could run training or do talks on various subjects, so much so that I have started my own company in this field and I love it. I have had some great experiences – and some pretty bad ones, which shape the values I hold now.

By way of a conclusion to this book, and in true military spirit, here are the qualities I've learnt to use as constants in my life:

- **Simplicity.** People will overcomplicate life for you very quickly so have a simple outlook and plan, to allow room for that to happen.

- **Example.** Live the example you want to set for yourself every day. Others will see this, so make no excuse for what you believe is right.

- **Calm.** Stay composed and don't flap. Calm is a superpower and in turn gives strength to others.

- **Integrity.** Be respectful, honest and dependable. Do what is right and not what is easy or convenient.

- **Communication.** Be clear in what you say and reinforce it often, making sure you really listen with your heart to what they say in turn.

- **Humour.** Don't take yourself too seriously and have a laugh when you can.

- **Decisions.** Make the hard call and once you do that, commit to your decision and follow it through.

- **Humility.** Know your own and your team's limitations and allow others the space to perform. You don't know everything so understand your importance is adjustable.

- **Empathy.** Tune into where people are in their heads and what they're facing; don't be afraid to give them a nudge.

► **Resilience.** Don't take things personally and be prepared for the worst while setting the conditions and hoping for the best. Be positive and understand things go wrong and you need to adapt in order to cope.

► **Confidence.** Be strong and assured in who you are and portray it without being brash.

I hope that you have taken something from all of this. Maybe you have an insight or have picked up something a little different. Maybe you've even fine-tuned your own factory-setting a little.

If I had known when starting off as a young soldier all the lessons I now understand, would I change how I did it if I got another shot at it? I don't think so. It wouldn't have been half as much craic if I had gotten everything right the first time around, would it?

GLOSSARY OF TERMS

2I/C: second in command

AAR: after-action review

Army Ranger Wing (ARW): the special operations force of the Irish Defence Forces. A branch of the Irish Army, it also selects personnel from the Naval Service and Air Corps.

asset: an asset is something like an aircraft or something you can use. Higher assets can be something up the food chain from you in rank or a HQ or other unit.

B6: a category of fully ballistic, bulletproof vehicles

BCD: buoyancy control device

blue on blue: friendly fire (inadvertent firing towards friendly forces)

bulling: a process of polishing boots to a mirror finish by using a cotton cloth

CO: commanding officer

colour sergeant: a senior non-commissioned officer rank in the British Army, just above the rank of sergeant

CP: close protection (another term for bodyguard duties)

CPO: chief petty officer

CS: company sergeant

CTD: command training depot

DA: direct action. Direct action mission to raid, assault, capture or neutralise a target or persons by forceful intervention.

DFAC: dining facility

DR: dispatch rider

DS: directing staff

DV: demand valve

evolution: in army terms, the process of being put through a rotation of a class, exercise or training event

FCA/ RDF: Reserve Defence Forces. The Reserve Defence Forces are the combined reserve components of the Irish Defence Forces.

flash: military insignia

garrison: a group of troops stationed in a fortress or barracks

GOC: general officer commanding a brigade

GPMG: general-purpose machine gun

granular training: specific training in the most finite detail, carried out to create an effective, instant reaction

Gurkhas: soldiers of Nepalese nationality native to South Asia recruited for the Indian and British Army

HEAT: hostile environment awareness training

HK: Heckler & Koch

HLS: helicopter landing site

HME: homemade explosive

HMG: heavy machine gun

HN: Haqqani Network

ICA: Israeli-controlled area

IED: improvised explosive device

infiltration: covert movement from a drop-off point to a mission target area

insertion: vehicular movement from a friendly base to a mission drop-off point

IS: Islamic State, an Islamist terrorist organisation also known as IS Khorasan or Daesh

Kosovo Force: a NATO-led international peacekeeping force in Kosovo

LUP: lying-up or staging position short of a target area where a team can stop and prepare, rest, admin kit and confirm exactly the plan to happen on target

LMG: light machine gun

LRCC: light recce commanders' course

LRRP: long-range reconnaissance patrol

LFTT: live-fire tactical training

medevac: a military helicopter or vehicle used for transporting wounded soldiers to hospital

MMG: medium machine gun

MOE: method of entry

MP: military police

NCO: non-commissioned officer. Non-commissioned officers usually obtain their position of authority by promotion through the enlisted ranks.

NGO: non-government organisation

NSDS: Naval Service Diving Section

NVG: night vision goggles

OP: observation post

OPS: operations

PF: pathfinder

principal: the person you protect on a close protection operation

protection operations: protecting an area, location or persons from attack

PSS: personnel support services (army counsellors)

PTI: physical training instructor

PUP: pick-up point

QRF (quick reaction force): an armed military unit on standby to rapidly respond to developing situations.

reorg: army-speak for reorganisation, an activity carried out to reset and get ready to move on

RPG: rocket-propelled grenade

RSM: regimental sergeant major

RTU: return to unit

SAF: small-arms fire

SF: Special Forces

SnR: surveillance and reconnaissance

SOF: Special Operations forces

SOP: standard operating procedure

SOTU: Special Operations task unit

SRCT: special route clearance team

tango: a terrorist

TI: thermal imaging

TL: team leader

tracer bullet: a bullet with a coloured incendiary charge attached to it so that it glows as it travels. It is used to identify the location of a target.

T-wall or Bremer wall: a 3.66m portable, steel-reinforced concrete blast wall of the type used for blast protection, used in several conflict zones

USSOCOM: US Special Operations Command

VBIED: vehicle-borne improvised explosive device

ACKNOWLEDGEMENTS

Thank you to my wife, Sinéad. For your patience, constant support and for putting up with me and all the madness, but mainly for just being you and for your ability to just get on with it. You are amazing and I love you.

To my boy Dan and the beautiful Louisa, thanks for having such a great impact on my life and for being such good fun to be with.

To my sisters and brothers, who covered my ass from the first day I had one to cover. Being the youngest, I was spoiled, but, truth be told, you spoiled me.

To Mike, my master, thank you for giving me a gift that I have had to fall back to all of my life. Fu Lung Chaun.

To Sarah, Aoibheann, Djinn, Laura, Teresa, Ellen, Claire and the Gill team for getting me to do this and for your words of wisdom throughout, thank you.

To the Defence Forces press office, thank you for your cooperation.

To Jamie D'Alton, who is the reason this book exists, I love that we always get there in the end. Thank you to all the team at *Ultimate Hell Week* for the effort you bring – to Anne, Real Jamie, Jane, Rosco, Jen, Sorcha and all the team.

To my fellow DS on the show – Ob, Ger, Staff, Rossa and John – thank you.

To my comrades in the Army and in Special Forces who taught me the ropes, backed me up, saved my ass and made me laugh. I can't name all of you or this will look like a phone book.

Thank you to Davy and Vino, for your kindness, to Dave R, Roger, Neil, Mr N and the team at Crean International Afghanistan and to the Roshan team, *tashakur*. To Aamir, Gus and the Pakistan team, my sincere thanks to all of you for guiding me.

A special thanks again to Rossa, for all your advice on this project and your support ever since we were dive buddies. It's hard to believe you were an officer once ...

To the lads in Cork, my lifelong friends, Keller, Foley, Trev, Bobby, Jonsey and Tom, thanks for listening to my crap all these years. At least now you have it in writing.

Finally, to my brothers from Golf One – Jeff, Staff, Bryan, Alex, Lorcan, Enda, Cathal and Baz – joined for life by an unbreakable bond.